FSC
www.fsc.org
MIX
Papier aus verantwortungsvollen Quellen
Paper from responsible sources
FSC® C105338

Alexandra Wasilewski

The Russian energy sector

The way for future development

Anchor Academic
Publishing

Wasilewski, Alexandra: The Russian energy sector: The way for future development, Hamburg, Anchor Academic Publishing 2015

Buch-ISBN: 978-3-95489-440-6
PDF-eBook-ISBN: 978-3-95489-940-1
Druck/Herstellung: Anchor Academic Publishing, Hamburg, 2015

Bibliografische Information der Deutschen Nationalbibliothek:
Die Deutsche Nationalbibliothek verzeichnet diese Publikation in der Deutschen Nationalbibliografie; detaillierte bibliografische Daten sind im Internet über http://dnb.d-nb.de abrufbar.

Bibliographical Information of the German National Library:
The German National Library lists this publication in the German National Bibliography. Detailed bibliographic data can be found at: http://dnb.d-nb.de

Hermannstal 119k, 22119 Hamburg
http://www.diplomica-verlag.de, Hamburg 2015
Printed in Germany

TABLE OF CONTENTS

LIST OF TABLES

LIST OF FIGURES

LIST OF ABBREVIATIONS

APS – Agreement of power supply

ATS – "Administrator of Trading System" JSC

BM – Balancing Market

CHP – Combine heat and power

CPP – Condensation power plant

CPSM – Competitive power selection model

DAM – Day ahead Market

EBRD – "European Bank for Reconstruction and Development"

EE – Energy efficiency

EIA – "International Energy Agency"

EIB – "The European Investment Bank"

ES – The Energy Strategy of Russia for the period up to 2030

ESE – "Energy systems of East" JSC

ETSAP – Energy Technology Systems Analysis Program

FER – Fuel and energy resources

FGC – Federal grid company "Federal grid company of Integrated Energy Systems" JSC

FGEF – "The French Government Environmental Fund"

FNPP – Floating Nuclear power plant

FTS – Federal Tariff Service

GDP – Gross domestic product

GHG – Greenhouse gasses

HVDC – High voltage direct current

HPP – Hydro power plant

IDGC – "Holding of inter-regional distribution grid companies" JSC

IES – Integrated energy systems

Inter RAO – "Inter RAO UES" CJSC

JSC – Joint stock company

LTPM – Long-term power market

MC – Commercial operator "Market Council"

PG – Power Grids

PE – Power and energy

PP – Power plant

PV – Photovoltaic

RA – Regulated agreements

RE – Renewable energy

RES – Renewable energy sources

RM – Retail market

SO – System Operator “System Operator of Integrated energy system” JSC

TGC – Territorial generation companies

TP – Technological platform

UNFCCC – United Nation Framework Convention and Climate Change

VEB – State Corporation "Bank for Development and Foreign Economic Affairs (Vnesheconombank)"

WB – The World Bank

WGC – Wholesale generation companies

WSM – Whole sale market

WWF – Worldwide Fund for Nature

ZFF – Zones of free flow

LIST OF UNITS

€/kW – euro per kilowatt

€/kW/year – euro per kilowatt per year

€/kWh – euro per kilowatt-hour

°C – Degree Celsius

Bn – Billion

GCal/h – Gigacalorie per hour

GW – Gigawatt-hour

GWh – Gigawatt-hour

Km – Kilometer

kV – Kilovolt

kW – Kilowatt

kWh – Kilowatt- hour

kWh/m^2 – Kilowatt-hour per square meter

kWh/GCal – Kilowatt-hour per gigacalorie

m^3 – Cubic meter

m – Million

MPa – Mega Pascal

Mtoe – Million ton of oil equivalent

MVA – Mega volt ampere

MW – Megawatt

m/s – Meter per second

Pcs – Pieces

PWh – Peta-watt-hour

Rub/y – Rubble per year

Toe – Ton of oil equivalent

Mtoe/y – Million ton of oil equivalent per year

TWh – Terawatt- hour

TWh/y – Terawatt-hour per year

USD – United States dollar

W – Watt

ACKNOWLEDGEMENT

I wish to express my sincere gratitude to my husband, to my mother and to my family for their assistance, love and support. I am also thankful for the support from organizations and institutions, such as the Katholischer Akademischer Ausländer Dienst (KAAD), the KfW Development Bank and the University of Flensburg, for providing a stable basis to fulfill and conclude the research, which now is being published in the shape of this book.

EXECUTIVE SUMMARY

For several years now, renewable energy technologies have been recognized as among the efficient sources for electricity production, playing an important part in the development of energy infrastructure in many countries without harming the environment. In addition, energy efficiency measures are becoming more highly prioritised in the renovation and sustainable development of the energy sector in most countries, due to the increasing amount of pollution from energy consumption and production.

This book represents an assessment of the current situation in the Russian electricity sector and the market potential for renewable energy and energy efficiency.

Over the past ten years, the Russian energy sector has changed significantly, following the restructuring of the largest energy holding, "RAO UES of Russia", which formerly comprised all parts of the energy sector. Today it is a big wholesale and retail market that provides a competitive environment, in which bidding for electricity and power contracts can take place, but still with high influence of governmental regulation and the tariff system.

One of the key objectives of this study is to analyse the following aspects: the institutional and legislative framework after the reconfiguration, the tariff system, the composition of the current energy matrix, the assessment of future electricity consumption and production, key government strategies and potential for renewable energy and energy efficiency development. The analysis gives an overview of the current situation of the sector, technical and economic potential of renewable energy development, and energy efficiency measures promoted by the government.

The second objective of the study is the analysis of the main entities of the market, the international institutions providing investment to the public sector of the market and potential areas of the market to promote renewable energy and energy efficiency, in order to provide the initial insight for possible investment into environmentally friendly projects.

The third objective of the study is to determine existing constraints preventing renewable energy and energy efficiency development, and measures to overcome these constraints for further sustainable development of the sector.

The institutional and legislative framework analysis showed that the government created a broad legislative framework for the sector reconfiguration and formation of an electricity market, still maintaining a high level of governmental control. Thus, after reconfiguration, the infrastructure sector, the electricity production by nuclear and hydro power plants, and the isolated sector from integrated energy systems (IES) of the country remain under government control. Although the main purpose of the reconfiguration was the creation of a competitive electricity market, the government still controls big parts of the tariff formation through the current legislative framework.

The renewable energy (RE) promotion was supported by the Federal Electricity Law (2003) with the introduction of a number of executive Acts. The Law and Acts established the first basis for RE development, but some very important additional executive Acts, such as premium price calculation, rules of costs calculation, grid connection subsidizing, etc. which were supposed to be adopted in 2010, have not yet been approved. This creates major barriers to RE facilities' entrance to the market and as a consequence for overall development of RE generation facilities.

The further development of RE facilities also highly depends on government energy strategies. The main strategy that determines RE and overall sector development is the "Energy strategy of Russia for the period up to 2030". The main emphasis of this strategy is on the development of CHP and condensate power plants, but slow development for RE facilities (4.5%).

The energy efficiency measures are also considered in the strategy. It is one of the key points for further promotion and development of the energy sector. The strategy aims at an overall reduction of energy consumption of 45%, where electricity sector consumption has to decrease by 30% by 2030. These measures predetermine values of the government program "Energy savings and energy efficiency improvement for the period till 2020". The public and private companies used this program as the basis for the development of their innovative and investment programs.

The comparison of the available RE economic potential in the sector and projected new supply in the sector showed that the country has enough RE resources to produce the required energy, but an analysis of the sector entities of the program showed that only public companies have a number of RE technologies to be developed in the near future, and this amount corresponds to a projected 4.5% by the government.

An analysis of the energy efficiency sector showed that it has huge potential for energy efficiency technologies in all spheres of electricity supply: generation, transmission and consumption. The private and government entities of the market are focused on conventional and innovative development of their assets. For instance, the main focus of the engineering industry is on highly efficient conventional technologies with low negative environmental impact, and with very low RE technology production.

The comparison of the necessary number of conventional power plants to be reconstructed and installed, with consideration of the possibility of installing a new generation of technologies, shows that the generation sector has high potential for significant improvement and with low negative environmental impact.

An analysis of the transmission sector showed that the main transmission company of the country "Federal grid company of Integrated Energy Systems" JSC implements governmental targets for further broad transmission system development and moreover is focused on Smart and Super grid implementation. It promises good potential for further efficient development of the transmission system.

The energy efficiency program in the consumption sector is under strong supervision by the government through a Technological Platform created by the Ministry of Energy. It is too early to investigate results of the energy efficiency program in the consumption sector, but after completion of the first phase of the program in 2015, the analysis of the results can be provided.

Investigation of existing constraints for RE and energy efficiency development showed that the highest constraint in the sector is the lack of a legislative framework. For renewable energy technologies, one of the important constraints is low engineering industry development in terms of renewable energy technology production, and the lack of additional support from the government budget.

One of the important measures, which have to be put into practice by the government, is the implementation of financial incentives for broad renewable energy development on a regional level.

In summary, according to all government programs, institutional and legislative frameworks, Russia has high potential for broad conventional energy efficiency development. This potential is supported by all entities of the market. The result of the country's chosen pathway will be seen at the end of 2030. But still, the country has areas where renewable energy solutions can be a more appropriate way for further development, and this sector has to be improved by the Russian government.

1 INTRODUCTION

Russia is the biggest country in the world by area, with a total of 17,075,200 km^2. It is situated in North Asia, bordering the Arctic Ocean, extending from Europe (the portion west of the Urals) to the North Pacific Ocean (CIA, 2012).

The population of the country is 138,082,178 people (July 2012 est.) with the growth rate in 2012 – 0.48%. Russia is a fast growing economy, with an annual GDP growth rate of 4.30% and a GDP per capita (PPP) of $16700 (CIA, 2012).

Russia is known as a country rich in natural resources, including major deposits of oil, natural gas, coal, and many strategic minerals; the energy sector in Russia is mainly based on conventional energy resources.

The composition of a Russian energy matrix is mostly represented by fossil fuels. According to the International Energy Agency in 2009, the highest primary energy supply was Natural Gas – 54%, in second place was Crude Oil with a contribution of 21%, in third place was Coal and Peat – 15%, and the consumption of Nuclear was 7%. Renewable energy resources including Hydro and Bio fuels were less than 5%. (IEA, 2011).

However, the share of fuel type for electricity production is different from primary energy supply. Power plants consume much less crude oil for electricity generation and more coal and peat. Nuclear consumption is much higher than in primary energy supply, as well as hydro resources. RE resources have a very small share in electricity generation.

Figures 1 and 2 represent the energy matrix in terms of primary energy supply and electricity generation by fuels.

The Russian energy sector is still going through big changes and improvement programs. The Russian public corporation of energy and electricity "Unified Energy System of Russia" – the RPC "UES of Russia" – was the main energy company in the Russian Federation before restructuring. It owned about 70% of Russia's installed electric capacity, 96% of its high-voltage grid and over 70% of its transmission lines. The reconstruction program started in 1998 and went through several stages during its implementation. The main target of reconstruction was the creation of competitive energy markets. This target was more or less reached in the middle of 2011, as planned (UES, RAO UES of Russia, 2005).

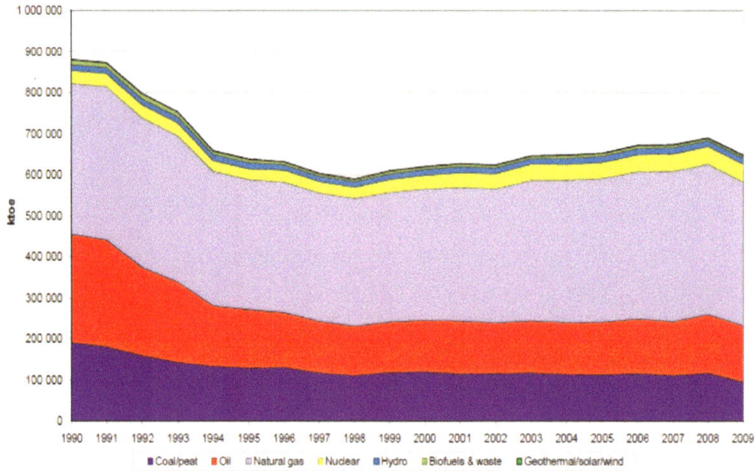

Figure 1: Total Primary energy supply, 1990-2009 (IEA, 2011)

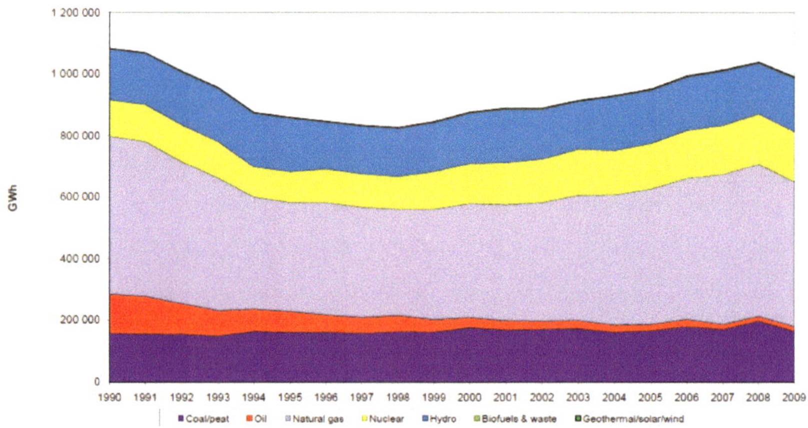

Figure 2: Electricity generation by fuel, 1990-2009 (IEA, 2011)

The Russian Federation is a party to the following international agreements (CIA, 2012):

- Air Pollution,
- Air Pollution-Nitrogen Oxides,
- Air Pollution-Sulfur 85,
- Biodiversity,
- Climate Change,
- Climate Change-Kyoto Protocol,
- Desertification,

- Environmental Modification,
- Hazardous Wastes,
- Ozone Layer Protection
- Etc.

Problem Statement

Considering the significant changes in the energy sector of the country, the high energy intensity and commitments of the country according to international agreements, it is relevant to investigate the energy sector of the country, to assess the current situation and to search for possibilities for renewable energy development.

Taking into account the abundance of conventional natural resources in the country, it is important to investigate and analyze energy efficiency programs as well as penetration of innovative smart technologies, in order to find a rational and environmentally friendly means of energy production and consumption.

Main objective

The main objective of this research is to analyze the Russian Energy Sector and to determine opportunities for renewable energy technologies and energy efficiency in electricity generation and transmission.

The specific objectives are:

1. To identify the government strategies for the energy sector;
2. To identify the legislative framework and tariff system;
3. To determine the composition of the energy matrix;
4. To assess installed and available capacities;
5. To identify main players in the Russian energy market (public and private entities);
6. To assess current and future energy demand and supply in the country;
7. To analyze the renewable energy potential in Russia;
8. To analyze the potential of energy efficiency measures in electricity generation and transmission;
9. To determine means of broader renewable energies development in the sector

10. To determine means of broader energy efficiency measures development in the sector.

Research Questions

The study will attempt to answer the following questions in detail:

1. What is the existing situation in the Russian energy sector?
2. What are the existing policies for renewable energy penetration?
3. What is the market potential for renewable energy development?
4. What is the market potential for energy efficiency generation, transmission and consumption?
5. Which market constraints prevent renewable energy and energy efficiency development?
6. How can existing constraints be eliminated, for broader renewable energy and energy efficiency development?
7. What policies would be required for further renewable energy and energy efficiency development?

Research methodology

The research perspective of the study is going to be a combination of quantitative and qualitative perspectives. The qualitative perspective will dominate during the research in order to best understand the legal, regulatory and institutional framework, and to estimate the dependence of the energy sector entities on the constructed framework and policy system. The quantitative approach will be used to collect and analyze data of energy demand and supply in the sector, existence and amount of conventional and renewable energy resources, to estimate energy efficiency measures and calculate replacement of conventional technologies by renewable, and to estimate potential for energy efficiency.

Regarding the research type, this study has descriptive, evaluative, constructive and prescriptive components, based on desk study.

The research objective will be approached by the following research process:

1. A collection of data to assess the current situation in the sector and estimate potential for renewable energy and energy efficiency development.

2. Identification of the present situation in the sector, renewable energy potential and potential of energy efficiency.
3. Evaluation of the possibility to exchange conventional energy resources with renewable energy sources.
4. Estimation of possibilities of energy efficiency implementation and comparison with projected implementation.
5. Evaluation of impact of governmental policy on renewable energy penetration.
6. Assessment of government and private entities feedback on existing policies.
7. Analysis, presentation and recommendations of further sector development.

Scope and limitations of the study

This study assesses the existing situation in the electricity sector of Russia, in order to estimate results of the completed restructuring program of the sector which was completed in 2011, to estimate the established situation for renewable energy technology penetration in the sector and possibilities for energy efficiency measure implementation.

The study first evaluates the new institutional framework of the system, legislative framework of the sector and tariff system, governmental strategies and programs, formation of electricity demand and supply. Then it describes the renewable energy potential in the country and energy efficiency measures, and contrasts proposed renewable energy penetration and energy efficiency measures by the government with available resources in the market for penetration. The study ends by identifying existing constraints in the sector for renewable energy penetration and energy efficiency measures and how identified constraints could be eliminated.

The study is limited by the electricity part of the market and does not include the heat part of the market. The research embraces the whole country, but does not consider regions and separate territories in detail. All financial figures in the study, which were found in Roubles and needed to be transferred in Euro, were calculated according to the currency rate on 01.10.2010: 1 Euro = 39.6028 Roubles (ProFinance, 2012).

The research is also limited in terms of environmental impact calculations, due to the lack of data. It suggests further investigation to address some of the issues that were not within the scope of this book.

2 OVERVIEW OF THE RUSSIAN ENERGY SECTOR

The second chapter of the book includes an analysis of the framework of the energy sector, assessment of electricity demand and supply, and renewable energy and energy efficiency potential.

2.1 Framework of the energy sector

Analysis of the framework of the energy sector embraces assessment of the institutional and legislative framework, tariff system and overview of government strategies of energy sector development, where the legislative framework and tariff system includes assessment of renewable energy legislation; assessment of government strategies is mainly focused on the main strategy of the sector and energy efficiency program.

Institutional framework of the sector

The Russian energy sector has significantly changed during the past 10 years, due to a reconfiguration program of the energy sector initiated by the Russian Government. The reconfiguration program which deals with transforming the energy sector of the Russian Federation under Government Resolution No. 526 began on 11th of July 2003 and was completed in the beginning of 2011.

The main reason for restructuring the energy sector was to create a competitive market in the sector and attract new investment. Based on this goal the sector is divided into three parts, where the government has 75%, 50% and less than 50 % of the share – in the different areas of the competitive sphere. Among the three sectors; generation, transmission and distribution, the generation sector is the most competitive sphere while the government is more involved in transmission and distribution.

The reconfiguration resulted in a general framework of the energy sector as represented in figure 3.

And as can be seen in figure 4, the market is represented by two layers (wholesale and retail market) and operated by market infrastructure.

Government		
More than 75% share	**More than 50% share**	**Less than 50% share**
System Operator (SO) "System Operator of Integrated energy system" JSC ("SO of IES" JSC)	*Holding of inter-regional distribution companies "Holding IRDC" JSC.*	*6 Heat Wholesale generation companies (WGC), 14 Territorial generation companies (TGC)*
Federal grid company "Federal grid company of Integrated Energy Systems" JSC ("FGC of IES" JSC)	*Hydro Wholesale Generation Companies (WGC) "RusHydro" JSC*	*Commercial operator "Market Council"*
Nuclear power stations "Rosenergoatom"	*Fareast energy company "Energy systems of East" RAO*	*Distribution companies, Service and repairing companies*

Figure 3: Russian energy sector after restructuring[1]

A set of additional laws and regulations were issued since 2011 and the market has the following structure:

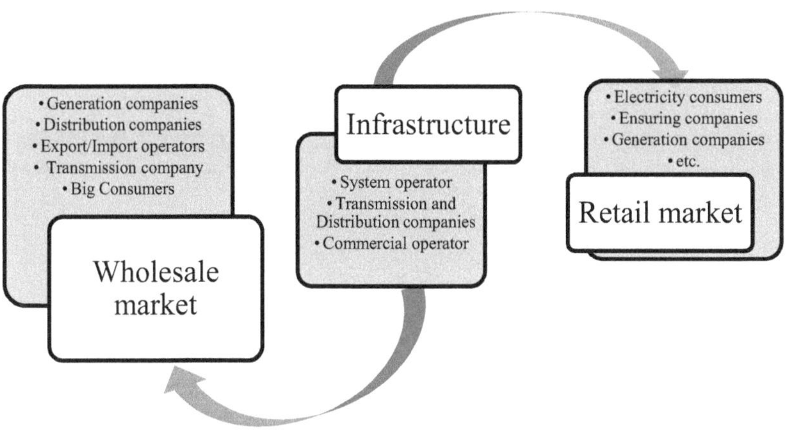

Figure 4: Electricity market structure[2]

[1] Derived by the author, based on consolidated information from www.np-sr.ru, www.rao-ees.ru

[2] Derived by the author, based on consolidated information from www.np-sr.ru, www.atsenergo.ru

The wholesale market is an energy and power market, where the market players include generation companies, with an installed capacity equal to or more than 25 MW; electricity export/import operators; distribution companies; large-scale consumers and transmission companies (in terms of buying electricity to cover losses during transmission) (Council, 2011)

Most of the country's generation assets are owned by one of the following companies (Council, 2011):

- 6 Heat wholesale generation companies
- Hydro generation company "Rushydro" JSC
- 14 territorial generation companies
- State Concern "Rosenergoatom"
- Far East energy company "Energy systems of East" RAO
- "Inter RAO UES" JSC

The main players of ***the retail market*** are electricity consumers; generation companies, with installed capacity of less than 25 MW; ensuring companies (distribution companies, which are obliged to supply electricity to any applied entity); companies of municipal service, which consume electricity to provide their services; system operator and other entities of dispatch management in isolated systems (Council, 2011).

Market Infrastructure is one of the important facilities of the market. It is divided into two parts: transmission and commercial infrastructure.

The transmission infrastructure is represented by transmission companies, which provide two main services (Council, 2011):

- Electricity transmission
- Technological connection between generation companies and power receiving consumers

These services are a natural monopoly and are controlled by the government.

There are three main companies of transmission infrastructure (Council, 2011):

- System Operator (SO) "System Operator of Integrated energy system" JSC ("SO of IES" JSC)
- Holding of inter-regional distribution companies "Holding IRDC" JSC
- Federal grid company "Federal grid company of Integrated Energy Systems" JSC. ("FGC of IES" JSC)

The most important player is the ***System Operator (SO)***, entitled to provide dispatch management and it has rights to give commands to any entity of the market, if it is necessary to maintain system stability and security (Council, 2011).

The *commercial infrastructure* is represented by the commercial operator – a non-commercial partnership "Market Council". The Market Council (MC) provides management of the wholesale and retail market through affiliated companies: "Administrator of Trading System" JSC and "Centre of Financial Transactions" JSC. The main functions of the Market Council are to ensure the operation of the market, to maintain a registry of the market and to develop and sign contracts of accession to the trading system of the wholesale market (Council, 2011).

Legislative framework and tariff system

To establish a competitive market in the sector, the government formulated a wide legislative framework for market regulation, which is represented as follows:

1. Federal law № 35- FZ "Electricity law" from 26.03.2003
2. Regulation of Russian Federation Government № 643 from 24.10.2003 "Working rules of wholesale energy and power market"
3. Federal law № 36- FZ "About the work features of the energy sector during the transition period"
4. Regulation of Russian Federation Government № 530 from 31.08.2006 "Approval of the rules of retail electricity markets in the transition period of reforming the energy sector"
5. Federal law № 41- FZ "About governmental regulation of electricity and heat tariffs in Russian Federation" from 14.04.1995
6. Regulation of Russian Federation Government № 576 from 14.07.2011 "About changes in paragraph 7 of rules of governmental regulations and utilization of tariffs for electricity and heat power in the Russian Federation"
7. Regulation of Russian Federation Government № 1172 from 27.12.2010 "About approval rules of the wholesale market of energy and power and about changes in some statements of Russian Federation Government in the questions of the work of the wholesale market of energy and power"

8. Regulation of Russian Federation Government № 813 from 06.10.2011 "About changes in the rules of the wholesale market of energy and power"
9. Regulation of Russian Federation Government № 238 from 13.04.2010 "Rules of determination of power prices on the wholesale electricity market"
10. Regulation of Russian Federation Government № 89 from 24.02.2010 "Rules of the wholesale market for the long-term power market"
11. Regulation of Russian Federation Government № 1178 from 29.12.2011 "About price formation in the area of regulated prices in the energy sector"
12. Regulation of Russian Federation Government № 1179 from 29.12.2011 "About identification and application by ensured suppliers' prices for electricity (power)"
13. Federal law № 401- FZ from 28.12.2010, Federal law № 33- FZ from 08.03.2011, Federal law № 394- FZ from 06.12.2011 "About changes in Federal Law *Electricity law*"
14. Decree of Federal Tariff Service № 228-e from 30.03.2012 "About approval of guidelines of tariff regulation with application of the return of investment method"
15. Decree of Federal Tariff Service № 302-e/3 from 29.11.2011 "About the approval tariff for services of the commercial operator, which provides by *ATS* Ltd"

It is based on this framework that the wholesale and the retail market function and the tariff system is worked out.

Thus, two price zones and one non-price zone were determined in the markets (Council, 2011):

- *The first price zone* includes a territory of the European part of Russia and Ural
- *The second price zone* includes a territory of Siberia
- *The non-price zone* includes the Arkhangelsk and Kaliningrad region, the Komi Republic and regions of the Far East. The existence of the non-price zone can be explained by the technological limitations of the sector; in general because the zone is not included within the integrated energy system of the country.

The tariff features of the non-price zone and price zone are described below.

To determine formation of the market prices it is necessary to consider the structure of each market.

The Wholesale market is characterized by the electricity and power market. *The electricity market* is comprised of different regulated agreements in the "day ahead" market (DAM) and balancing market (BM):

The sector of regulated agreements (RA) from 2011 functions only in the sector of price zones. These agreements are signed up only for selling electricity to private households or to those consumers who can be determined as "private households" and for inter-regional transmission companies of North Caucasus. (Council, 2011)

The amount of electricity (power) and prices for RA are formulated according to Regulation № 1178 from 29.12.2011. By this Regulation the Federal Tariff Service (FTS) prepares a consolidated forecast balance, where every producer of electricity supplies not more than 35% of its overall amount of electricity (power) to the market (Council, 2011).

The amount of electricity (power), which is not supplied by RA, is supplied by agreements in the "day ahead" market and balancing market (Council, 2011).

The MC is responsible for DAM. It collects price bids from sellers and buyers the day before real supply of electricity, and makes competitive selection, by choosing margin price. Those who apply with the lower price are chosen first to supply the demand. (Council, 2011)

The RA and DAM are compensated by BM. Three hours before real supply, the System Operator forms the final balance of demand and supply to secure the stable operation of the system and to compensate a lack of supply or oversupply. The penalty or premium for all parties of the market is calculated according to the lack of supply or oversupply or the exact fulfilment of obligations. (Council, 2011)

The power market

The main purpose of the power market is not only to supply the required power, but also to collect funds for the further development of the energy sector.

From 1st June 2008 there is a functioning competitive power selection model (CPSM). The principle of CPSM is the same as DAM. Within this model there are two types of agreements: Long-term power market and the Agreement of power supply (Council, 2011):

- ✓ ***Long-term power market.*** Under Regulation № 89 from 24.02.2010 the power market from 2010 operates by a new model – the Long-term power market (LTPM), whereby the RA since 2011 supplies only the population and consumers who can be determined as the population. LTPM is functioning by CPSM, and this selection is provided for 4 years.

  Power price in this type is formed according to zones of free flow (ZFF)[3] or to the zone of limited competition, as it is determined by the Federal Antimonopoly Service. Russian Federation Government determines price limits for ZFF. Only 3 ZFF from 27 are not limited by the Government – ZFF Centre, Ural, and Siberia. There are functioning market selection according to Regulation № 813 from 06.10.2011.
- ✓ ***Agreement of power supply.*** Generation companies can supply power by signing "Agreement of power supply" (APS). By this agreement heat power plants get guaranteed payment for 10 years, nuclear and hydro power plants get guaranteed payment for 15 years. Rules for the calculation of the power prices according to APS are determined by the Regulation № 238 from 13.04.2010

Generation Companies have the right to choose the type of agreement by which to sell the power to the market (Council, 2011).

The Retail Market is the market of electricity and power that has been bought in the wholesale market, and also electricity from generation companies with an installed capacity of less than 25 MW. The entities of the retail market were considered in the chapter *Institutional framework of the sector.*

Price formation in the retail market, as in the wholesale market is determined according to price zones, non-price zone and consumers who are determined as private households.

Domestic customers and entities which form the market from non-price zones are reimbursed for electricity and power by regulated prices. (Council, 2011)

Regulated prices for domestic customers are established yearly by FTS, according to the Regulation of the Russian Federation Government № 1178 from 29.12.2011. FTS

---

[3] "Zone of free flow - part of the Integrated electricity system , where electricity (power) produced or planned to be supplied by generation equipment can be exchanged by electricity (power) produced or planned to be supplied by other generation equipment, which has the same technical characteristics in the zone of free flow." - "Market Council", Glossary, http://www.np-sr.ru/presscenter/glossary/SR_0V005278

establishes maximum and minimum limited prices for each region of the Russian Federation, as stated in the forecast of consolidated balance (Council, 2011).

According to the prices regulated by FTS the Regional Tariff Services establish regulated tariffs for consumers (Council, 2011).

Regulated prices for other consumers in a non-price zone are calculated in two parts (Council, 2011):

1. Shifting price on the wholesale market up to final consumers: it is a sum of the payments, that include shifted price after the bids on the wholesale market and reimbursement of the difference in consumption (over or under consumption) paid by the retail consumer.

   The difference in the consumption can be provided by retail consumers only if the consumer's connected capacity is more than 750 kVA. Such consumers have to plan their own consumption on an hourly basis. Those consumers who have a capacity of less than 750 kVA are under the control of the infrastructure operators.

2. Payment for the electricity transmission: this payment is composed of the infrastructure service reimbursement and the distribution price of the ensured supplier.

Non-regulated prices are implemented on the retail market in the price zones of the wholesale market. In general, the range of non-regulated prices in the retail market is composed of the following parts (Council, 2011):

Average non-regulated price of electricity (power) on wholesale market

(determined by "ATS" Ltd.)

+

Infrastructure service reimbursement + distribution price of ensured supplier

+

Reimbursement of commercial operators

The distribution price for ensured suppliers is calculated and published by the Regional Tariff Services in each Region of the Russian Federation.

The next step is to consider the components of the tariff for electricity transmission.

The players in the **market infrastructure** for transmitting electricity were considered above. The tariff system of the market infrastructure is composed of two components: the tariff for *transmission infrastructure* and the tariff for commercial infrastructure (Council, 2011):

1. The final consumer pays the following tariff for "transmission operators":

   a. *Cascade principle.* Required gross income of high voltage transmission companies + required gross income of middle voltage transmission companies + required gross income of low voltage transmission companies = Transmission (distribution) tariff

   b. *The principle of uniform tariffs.* Tariff in one voltage class in one region is equal for all consumers of this region, even where different grid companies operate.

   The Federal Tariff Service by the decree of FTS № 228-e from 30.03.2012 establishes the value of required gross income, by description of methods of required gross income calculation, with the consideration of return of investment.

2. Commercial operator reimbursement according to established tariff by FTS. The value of the commercial operator tariff is published on the official web-page of FTS by the decree of FTS № 302-э/3 from 29.11.2011.

In summary, the Energy sector tariff system is represented by the figure 5.

The tariff values, which are calculated by the Regulations of the Government of the Russian Federation and Decrees of the Federal Tariff Service, include the amount of income, which is necessary to return invested capital. The FTS ensures that the considered amount is enough for new investment of the companies. (Service, 2012, p. 3) The federal grid company confirms affirmations of FTS, at the same time the Far East energy company complains about insufficient income. (FGC, 2012, p. 2) (ESE, 2009, p. 4)

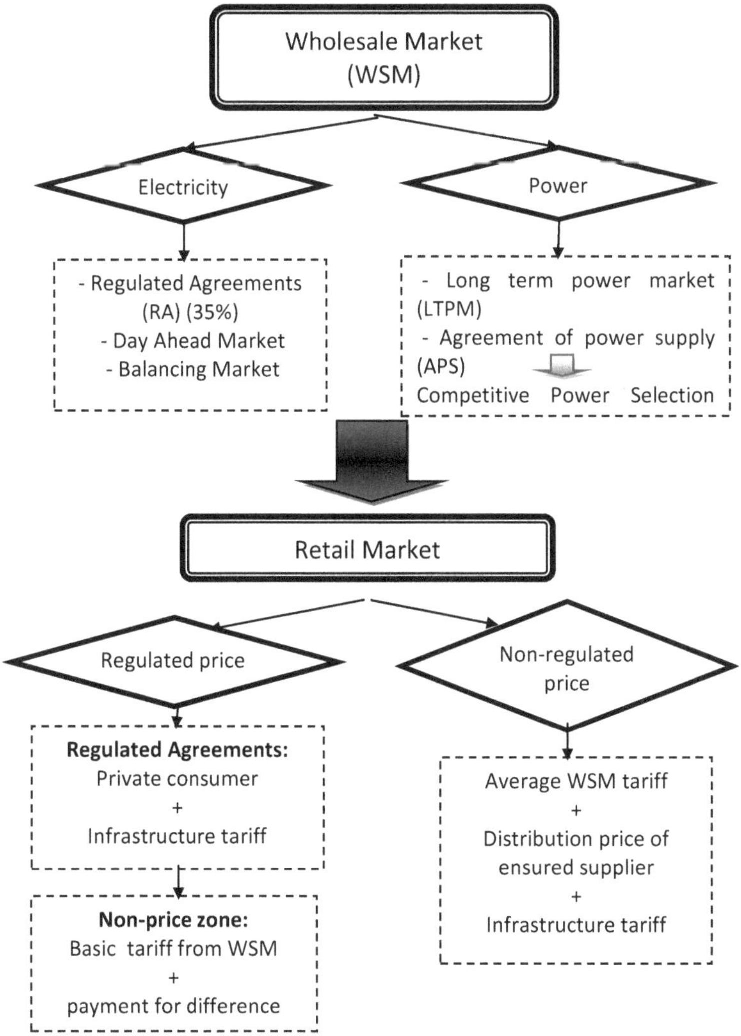

Figure 5: Tariff system in the Russian energy sector[4]

[4] Derived by the author, based on consolidated information from www.np-sr.ru, www.atsenergo.ru, http://www.fstrf.ru/

Renewable Energy legislation framework and tariff system

The Federal law № 35- FZ "Electricity law" from 26.03.2003 (with amendments № 250 FZ from 4.11.2007, № 401 FZ from 28.12.2010) establishes the main principles of Renewable Energy (RE) development in the market.

As stated in the Federal Law the following executive acts were introduced:

1. Regulation of the RF Government № 426 from 03.06.2008 "On the qualification of a renewable energy generating facility". According to that regulation, the RE generating facility has to be qualified by the following criteria:

    - Generation is based on renewable energy resources or a combination of RE sources and other sources
    - Generator has to be in service (being in service, not in a repairing regime, or out of service)
    - Generator is connected to the grid system and equipped by metering devices.

    There are no special rules for isolated systems. These will probably be classified as the conventional isolated system.

2. Regulation of the RF Government № 850 from 20.10.2010 "Criteria of subsidizing grid connection of renewable energy generating facilities less than 25MW from federal budget".

    When the generator is approved by the required criteria, it can be subsidized for the connection to the grid.

3. Decree of RF Government № 1-p from 08.01.2009 "Main directions of government policy in the sphere of energy efficiency on the basis of RE sources utilization up to 2020"

    The decree considers the main rates of RE development (besides hydro power plants more than 25 MW)

    - 2010 – 1.5%
    - 2015 – 2.5%
    - 2020 – 4.5%

    The same rates are considered in the government strategy "Energy Strategy of Russia for the period up to 2030".

4. Decree of Ministry of Energy RF № 187 "This regulation deals with the procedure of maintaining the registry of issuance and redemption of certificates, confirming

volume of production of electricity on qualified generation facilities, and functioning on the basis of RE sources".

The Decree gives a definition of the certificate, and how this certificate has to be issued and paid.

The Market Council is responsible for: qualifying the RE generator to issue and pay certificates; developing and signing RE contracts of connection to trading system of the wholesale market; controlling the purchase of RE electricity on the market by consumers.

Nowadays there is no special tariff system for RE generators. There are several incentives which were supposed to be affirmed in 2010 by the Government of the Russian Federation according to Federal Electricity Law (Council, 2011):

1. Regulation of RE premium calculation, which has to be added to the equilibrium price of the market.
2. Regulation of the amount of obligatory purchase of electricity produced by RE generators.
3. Rules of the cost calculation of electricity produced by RE on the wholesale and retail markets, and in non-price zones
4. Rules of cooperation of RE generators with grid companies and ensured suppliers.
5. Criteria of subsidizing grid connection of RE generators with installed capacity less than 25 MW

The official Government sources do not publish exact information when these incentives are going to be affirmed. In general, the legislative framework for RE demands more detailed consideration by the Government. For the Market Council there are still many inexplicit moments in RE generator entrance to the market.

In consonance with the energy efficiency legislation, the RF Government issued regulation № 2446-p on 27.12.2010 entitled: "About approval of government program: Energy savings and increase of energy efficiency till 2020".

The overview of the program is going to be considered in the next chapter.

Overview of government strategies of energy sector development

The main Russian governmental strategy, which predetermines the development of the whole energy sector in Russia, is "Energy strategy of Russia for the period up to 2030". The strategy has been approved by decree of the Russian Federation Government №1715-r, from 13.11.2009.

The strategy is the main determining element for various strategies in the energy sector, such as:

- General scheme of gas sector development for the period to 2030
- General scheme of allocation of electricity facilities for the period to 2020
- General scheme of oil sector development for the period to 2020
- Government program "Energy savings and energy efficiency improvement for the period till 2020"
- Etc.

This sub-chapter provides an overview of the Energy strategy, and key points of the Energy savings program. The General scheme of allocation of electricity facilities is not going to be considered in this chapter, since it is highly interconnected with the Energy Strategy. The main targets of the General scheme of allocation of electricity facilities are highly dependent on the Energy strategy.

The Energy Strategy for a period up to 2030 (ES)

The main purpose of ES is the creation of an innovative and effective energy sector. The purpose is going to be reached by implementation of the following objectives of the ES:

- Creation of a stable institutional environment in the energy sector
- Modernization and creation of a new energy infrastructure
- Improvement of the energy and eco-efficiency of the Russian economy and energy sector
- Improvement of efficiency of fuel and energy resources (FER) production, mining and refining
- Further integration of the Russian energy system into the world energy system

There are three phases of ES implementation:

1st phase – until 2013-2015

2nd phase – until 2020-2022

3rd phase – until 2030

The main concepts of the ES will be implemented in accordance with these three phases.

The first concept is *to ensure energy security* for the country. It includes the following indicators, whose implementation will provide energy security.

The indicators are described in table1.

Table 1: Energy security indicators[5]

1 phase	2 phase	3 phase
Growth of energy consumption per capita, in % to the level of 2005		
at least 10	at least 20	at least 40
Growth of electricity consumption per capita, in % to the level of 2005		
at least 13	at least 43	at least 85
Growth of engine fuel consumption per capita, in % to the level of 2005		
at least 23	at least 41	at least 70
Reduction of average level of deterioration of capital funds, %		
10	10	5
Maintaining a stable reserve of electricity and heat capacities: Electricity power reserve on the level of 17% Heat supply reserve on the level of standard reserve		

Considered consumption growth means that the fuel and energy complex of the country has to guarantee supply on the level considered in each phase of consumption. The comparison of electricity consumption indicators and demand and supply forecast values, considered in chapter 2.2., show that predicted growth of electricity demand in phase 1 was correct and realistic.

The second concept is *to ensure energy efficiency of the economy*. The indicators of the second concept are described in the table 2.

Table 2: Energy efficiency of economy indicators[6]

1 phase	2 phase	3 phase
GDP energy intensity, in % compared to 2005		
No more than 80	No more than 62	No more than 46
Creation of additional energy potential of economic development		
100 Mtoe/y	200 Mtoe/y	300 Mtoe/y
Expenditure formation of high technological energy service sector, in amount of		
200 bn rub/y	300 bn rub/y	400 bn rub/y
Reduction of losses and self-consumption of fuel and energy complex, %		
at least 1	at least 1	at least 0.5
Reduction of fuel consumption for heat production by boiler plants and heat PP, %		
at least 2	at least 6	at least 10

5 Derived by the author, based on information from "Energy strategy of Russia, for a period up to 2030"

6 Derived by the author, based on information from "Energy strategy of Russia, for a period up to 2030"

The third concept is *to provide environmental energy security*. As seen in table 3, at the end of the second phase in 2022, the energy sector has to have decreased the level of air pollution, water pollution and waste creation by a minimum of 30%; and in 2030 this level has to reach a 50% reduction. Also, it is planned that the level of greenhouse gases (GHG) emission in phase two will stay below the level of 1990 – maximum 92%, and in 2030 will reach the level of 105%.

Table 3: Environmental energy security indicators[7]

1 phase	2 phase	3 phase
Reduction of indicators of air pollution, water pollution and creation of wastes by energy sector		
at least 15%	at least 30%	at least 50%
Ensure level of GHG emission, in % to the level of 2005		
No more than 83	No more than 92	No more than 105
Coefficient of GHG disposal (in 2005 - 75%)		
95%	95%	95%

Figure 6 represents the predicted dynamics of CO_2 emission max/min level in percentages compared to the level of 1990[8].

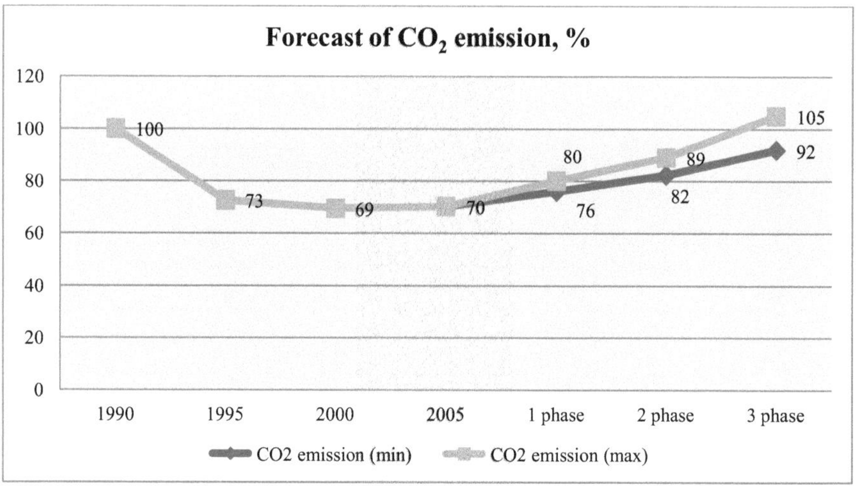

Figure 6: Forecast of CO_2 emission, %[9]

[7] Derived by the author, based on information from "Energy strategy of Russia, for a period up to 2030"

[8] Derived by the author, based on information from "Energy strategy of Russia, for a period up to 2030"

According to ES, the energy sector of the country will be developed with a high share of fossil fuel production and consumption due to the high level of fossil fuel resources in the country. Thus gas production in the country in 2005 had a share of 42.5% and oil – 38.6% from the total production of FER, the non-fossil fuel products took only 6.4%. In 2030 the level of non-fossil fuel resources will reach 8.3% and oil production will decrease to 32.3%. (Government, 2009, p. 93)

The share of overall production of FER is represented in figure 7.

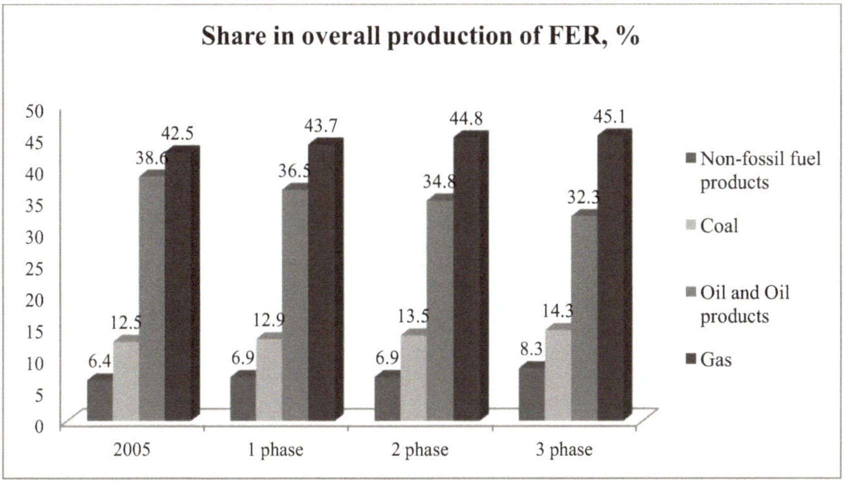

Figure 7: Share in overall production of FER[10]

The installed capacity of the electricity sector is going to increase by 102% of installed capacity in 2005 with an optimistic scenario, and by 66.7% with a moderate scenario by 2030. The highest increase of installed capacity, an increase of 141%, is going to be by Condensation Power Plants (CPP) in the optimistic scenario and 102% in the moderate scenario[11]. (Government, 2009, pp. 113-115)

The share of installed capacity of power plants in 2030 is going to change in comparison with 2005. Thus, the highest installed capacity in 2005 was CHP plants – 36%, and in 2030

---

[9] Derived by the author, based on information from "Energy strategy of Russia, for a period up to 2030"

[10] Derived by the author, based on information from "Energy strategy of Russia, for a period up to 2030"

[11] Condensation power plant is a kind of PP that has similar characteristics as CHP; it can produce electricity or combined electricity and heat, and use for the production fossil fuel resources.

condensation power plants are going to take the lead with a share of 35%. The share of Hydro PP is going to stay more or less the same; with an increase of 1%. The share of Nuclear PP and Renewable Energy plants will increase by 3% and 2% respectively. (Government, 2009, pp. 113-115)

Figure 8 shows the installed capacity by each phase and share of PP in the electricity sector in 2005 and 2030.

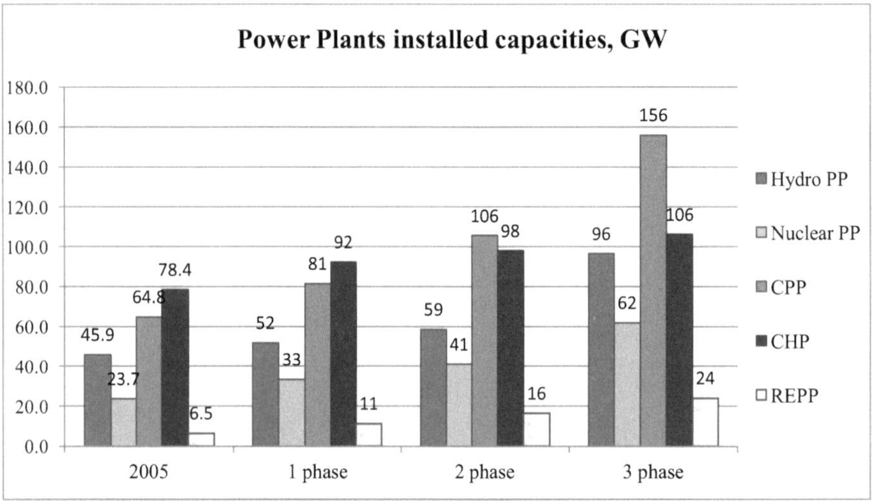

Figure 8: Power Plants installed capacities[12]

Renewable energy doesn't have such high priority development in non-fuel energy. It is considered by the program, which indicates that RE has to increase from 0.5% in 2005 to 4.5% in 2030. To reach this target there should be an installation of 23 to 33 GW of RE power plants in the sector (Government, 2009, p. 74). Figure 9 shows RE development during the three phases.

The energy strategy considers energy saving measures as well. The potential for energy savings is estimated at the level of 45% from current consumption of FER. Thus, consumption of energy resources can be decreased by 20% in the supply of heat, 30% in electricity sector, 40% in industry and transport and 50% in the residential sector. (Government, 2009, pp. 13-16)

As a result, the energy sector for further development from 2008 to 2030 will need 2697 bn USD dollars of 2007 prices, which include all types of industries, electricity sector and

[12] Derived by the author, based on information from "Energy strategy of Russia, for a period up to 2030"

energy savings. Figure 10 represents necessary investment in the sector (Government, 2009, p. 118).

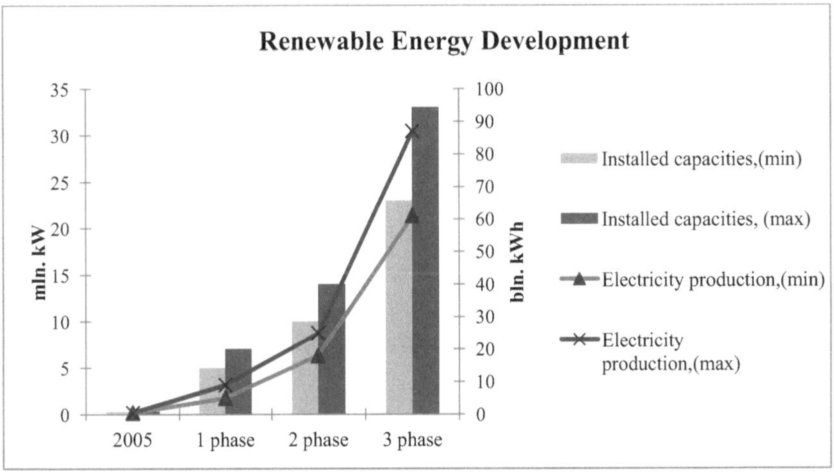

Figure 9: Renewable Energy Development[13]

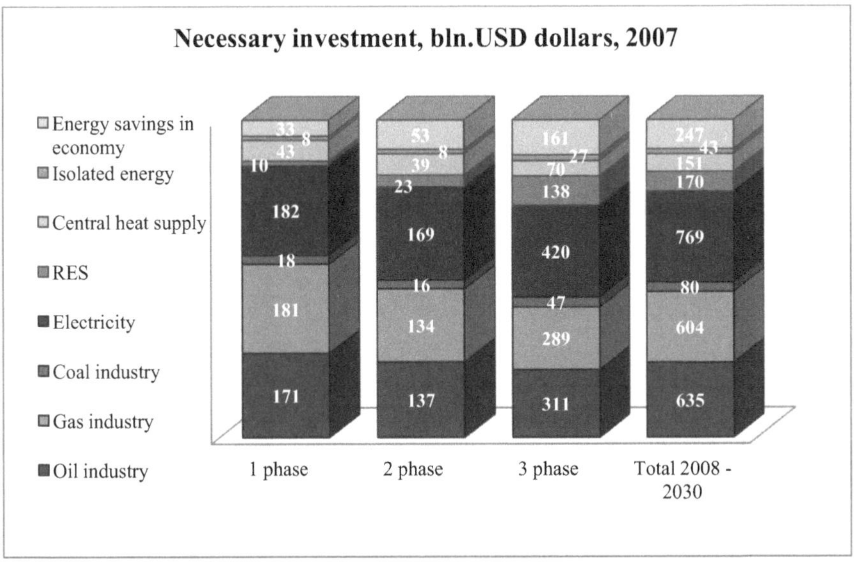

Figure 10: Necessary investment in the Energy Sector of Russia[14]

---

[13] Derived by the author, based on information from "Energy strategy of Russia, for a period up to 2030"

Government program "Energy savings and improvement of energy efficiency for the period up to 2020".

The energy savings program (ESP) was approved by the Russian Government statement № 2446-r on 27.12.2010. It contains 9 subprograms, which are energy savings and improvement of energy efficiency in the electricity sector, heating supply and systems of municipal infrastructure, industry, agriculture, transport, state institutions and service sector, housing sector, incentives in the Regions of the Russian Federation and methodological, informative and human resources support.

The main purpose of the program is to reduce GDP energy intensity by 13.5% for the period 2007-2020, to reach the annual primary energy savings of at least 100 Mtoe up to 2016 and 195 Mtoe up to 2020. (Government, 2010, p. 3)

The ESP has two phases of realization: 1st phase – 2011-2015 and 2nd phase – 2016-2020.

To implement ESP it is planned to attract investment from the Federal budget, from budgets of the Regions of RF and non-budget sources. Expected results of the program realization are represented in table 4.

Table 4:Expected results from Energy saving program[15]

Name	1 phase	2 phase
Conservation of primary energy, Mtoe	334	1124
Conservation of natural gas, bn m3	108	330
Conservation of electricity, bn kWh	218	630
Conservation of heat energy, m Gcal	500	1550
Conservation of oil products, m ton	5	17
Annual reduction of GHG, m ton CO_2 eqv.	673.5	2436

[14] Derived by the author, based on information from "Energy strategy of Russia, for a period up to 2030"

[15] Derived by the author from "Energy savings and improvement of energy efficiency for the period till 2020"

2.2 Assessment of demand and supply and future development of the sector

The electricity system of the country works under the control of the System Operator, which accomplishes the work of the system through 7 branches – integrated energy systems (IES) of Central, North-Western, Volga, Southern, Ural, Siberia and Far East. The Far East IES mostly works in an isolated regime (SO, 2012).

The forecast and formation of the future energy system of the country is the responsibility of SO with the support of FGC. Thus SO annually forms and corrects the demand and supply program for a period of 7 years (Governemnt, 2009).

Current and future electricity demand

Electricity consumption in 2010 was 988.96 bn kWh, which is higher than the level of 2009. The growth of electricity consumption can be characterised by the economic recovery during the period after the economic crisis.

A moderately optimistic economic growth was used to predict the demand growth of the nation in the same period. The main drivers were reintegration of production capacities before the crisis and growth of investment activity, especially of infrastructure projects. It was also considered that GDP growth rate is the main factor for the growth of electricity consumption on the levels of 3.9% to 4.5% in 2012, 3.5% in 2013 and an overall growth of 2.6% over the period 2011 to 2017. (SO, 2011, p. 2)

Figure 11 represents current and future electricity demand.

The highest electricity growth rate is predicted in the IES South with an increase of 3.71%. The main factors of this increase are (SO, 2011, pp. 3-6):

- Upgrade of manufacturing production
- Construction of agriculture, electrical and transport machinery.
- Development of the steel industry
- Oil and gas mining in the Caspian shelf
- Development of pipe line systems (Russian Caspian pipeline consortium)
- Construction of Olympic places
- Development of Rostov, Volgograd, Krasnodar trade, logistic and financial centres of national and regional level
- Development of the agricultural sector
- Tourism and recreation

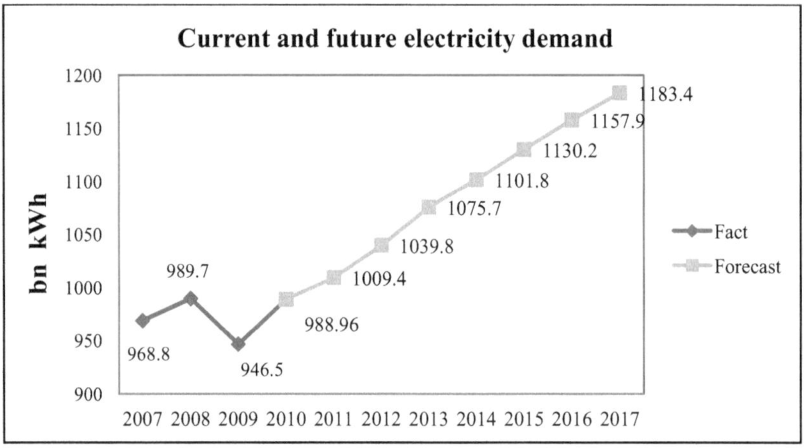

Figure 11: Current and future electricity demand[16]

IES East is in second place with an average growth of 3.64%. The main reasons for the growth are: the construction of the second part of the oil pump system in Amur, Khabarovsk and Primore regions; construction of the oil-chemical complex in the Nachodka area; development of the Elgin coal ore with railway construction in Ulak-Elga, steel plants in Priamurie; construction of infrastructure for the Asia Pacific Economical Cooperation Summit; construction of the University campus in Russian Island, etc. (SO, 2011, pp. 6-8)

IES Center stands in third place with a growth of 3%. The main reasons for the growth are further development of Moscow and the Moscow region and high investment projects, such as: construction of research and production electricity steel plants, construction of cement plants, creation of technological park "Lyudinovo", integrated system of transport communication and development of research-innovative centres. (SO, 2011, pp. 6-8)

Consumption growth of IES Siberia is 2.73% which has been shown by construction of two aluminium plants, the Siberian electric steel plant, capacity expansion in the Irkutsk aluminium plant, expansion of the "East Siberia – the Pacific" oil pipeline, infrastructure development and civil buildings construction in Krasnoyarsk. (SO, 2011, pp. 6-8)

The growth rate in the remaining three IES: Volga, North-West and Ural is 2.33%, 2.12% and 1.86% respectively. The main characteristic of growth in North-West is further development of the St. Petersburg and Leningrad region; for Volga IES it is the development of the Republic Tatarstan and the Samara and Nizhnii Novgorod regions. The lowest growth

[16] Scheme and program of IES of Russia development for the period 2011-2017, from 29.08.2011

in the Ural region is characterized by further development of the Sverdlov and Chelyabinsk regions, as well as oil mining and oil-gas chemical production. (SO, 2011, pp. 6-8)

The SO uses power consumption predictions in the sector, for further predictions of necessary power capacity development. With consideration of the absolute value of the power reserve available and the forecasted demand growth rate, the system is expected to continue working until 2015 with a complete stability. Power demand growth is represented in the Figure 12.

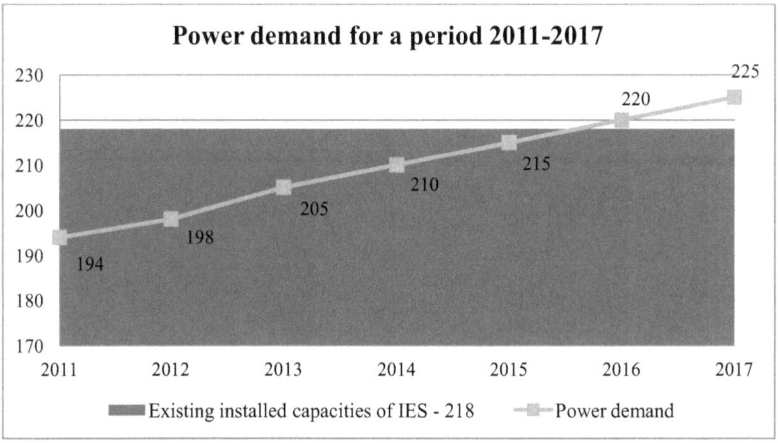

Figure 12: Power demand for a period 2011-2017[17]

Current and future electricity supply

Currently Russian Electricity supply is represented by 600 power plants with a total installed capacity of 218.2 GW. In 2011 an additional 4.7 GW of new capacities was installed and old plants with capacities of 1.6 GW were decommissioned (Federation, 2012).

The high voltage grid system is represented by 10,200 lines of 110-115 kV. The Russian electricity system works in synchrony with the energy systems of Azerbaijan, Belarus, Georgia, Kazakhstan, Latvia, Lithuania, Moldova, Mongolia, Ukraine, and Estonia. The Kazakhstan grid line accomplishes synchrony with Kyrgyzstan and Uzbekistan. Through that synchrony line the Russian electricity system also exports electricity to China. Through Vyborg transforming complex in St. Petersburg it exports electricity to some parts of Norway. (SO, 2012).

---

[17] Scheme and program of IES of Russia development for the period 2011-2017, from 29.08.2011

The SO predict installed capacities for 7 years (last prediction 2011-2017) with the consideration of installed capacities in the country, commissioning of new capacities, modernization and reconstruction or decommissioning of old capacities (SO, 2012).

Thus it is planned for the period of 2011-2017 to decommission 7.83 GW of installed capacities, including 0.83 GW of nuclear PP. Figure 13 represents capacities, which are planned to be decommissioned during this period (SO, 2011, p. 27).

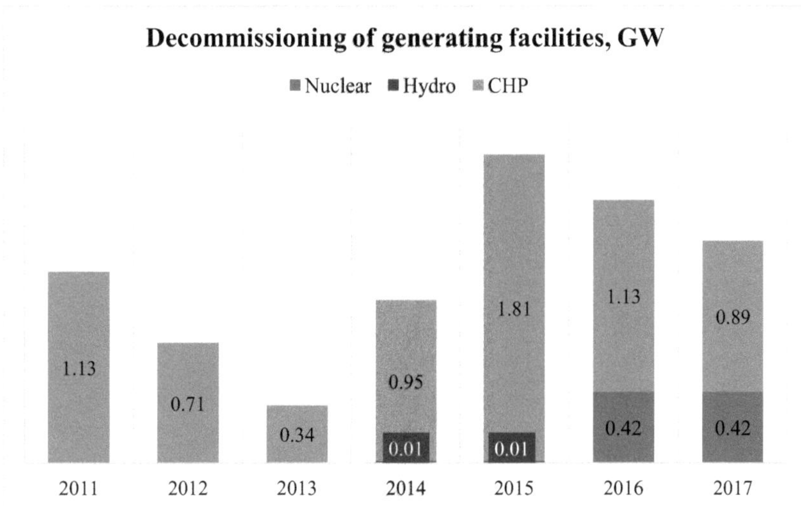

Figure 13: Decommissioning of generating facilities[18]

To satisfy electricity demand it is necessary to install 50.05 GW of new capacities during the period, with 9.88 GW of Nuclear, 4.09 GW of hydro, 0.98 GW of storage hydro, 34.44 GW CHP and 0.66 GW RE power plants (SO, 2011, p. 34).

The current system of the country, after decommissioning of old installed capacities during the period is illustrated in figure 14.

[18] Scheme and program of IES of Russia development for the period 2011-2017, from 29.08.2011

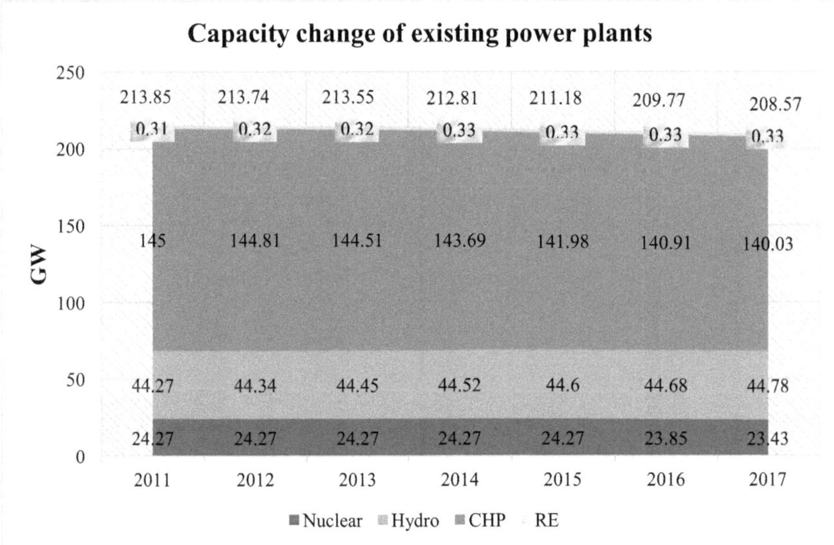

Figure 14: Capacity change of existing power plants[19]

The System Operator considered probability of necessary installations according to the following gradation (SO, 2011, pp. 34,35):

1. Generation facilities, where construction is planned according to liabilities by APS.
2. Generation facilities, which are included in investment programs of “Rosenergoatom Holding” Ltd., “Rushydro” Ltd., Fareast energy company "Energy systems of East" RAO, etc.
3. Generation facilities, which are include in the considered balance by FTS.
4. Generation facilities, which already have agreements for technological connection.
5. Generation facilities, which are included in the program of improvement of reliability of Moscow and the Moscow region energy systems
6. Generation facilities with high probability of construction.

Volumes and structure of commissioning capacities are represented in figures 15 and 16.

---

[19] Scheme and program of IES of Russia development for the period 2011-2017, from 29.08.2011

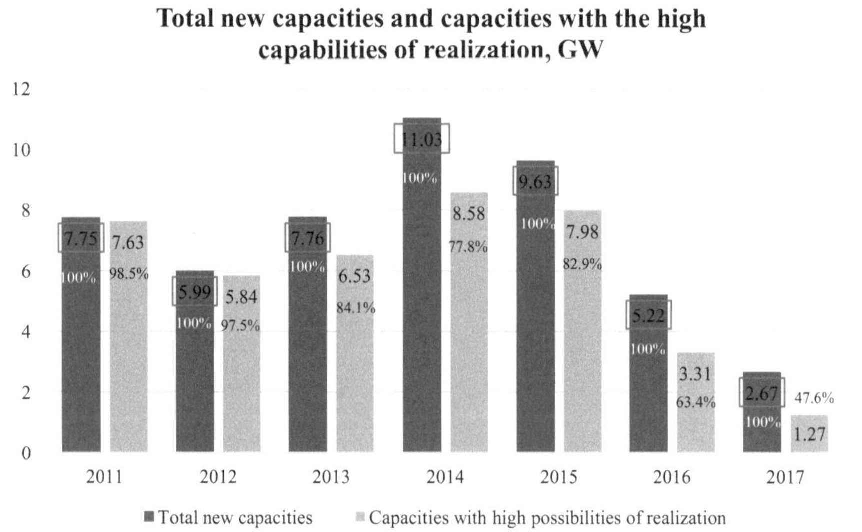

Figure 15: Total new capacities with consideration of capacities with high capabilities of realization[20]

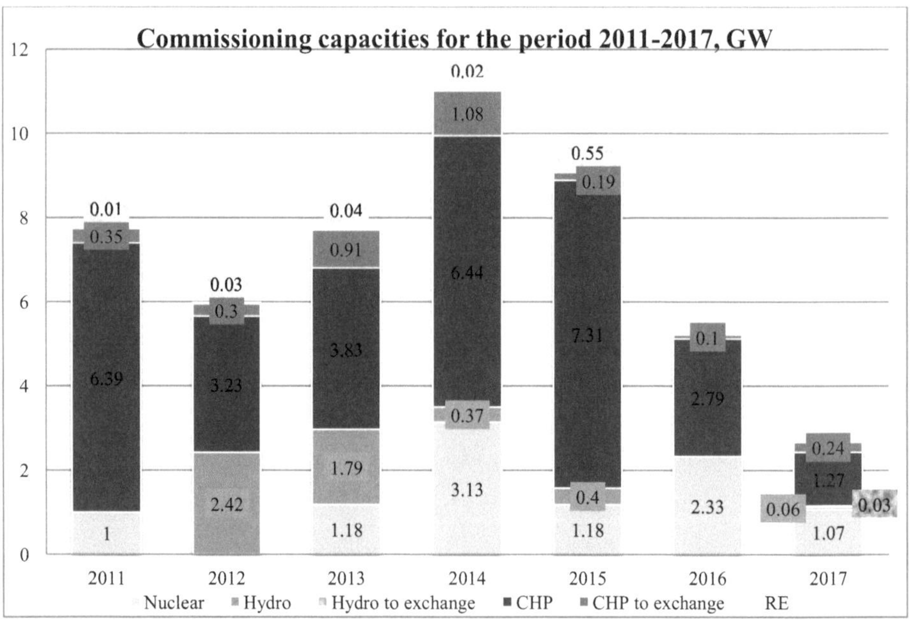

Figure 16: Commissioning capacities for the period 2011-2017[21]

---

[20] Scheme and program of IES of Russia development for the period 2011-2017, from 29.08.2011

Commissioning of new capacities of nuclear power plants is most likely to involve the installation of new generating units in the existing power plants. Hydro power plants considered for commissioning are those which are under construction. Growth of CHP will be enhanced by the installation of innovative steam-gas power plants, whereas the installation of wind and hydro PPs, with a range of capacities from 12 MW to 500 MW is expected to contribute to RE growth (SO, 2011, pp. 39-45). The structure of the system during the period of 2011-2017 will be as represented in figure 17:

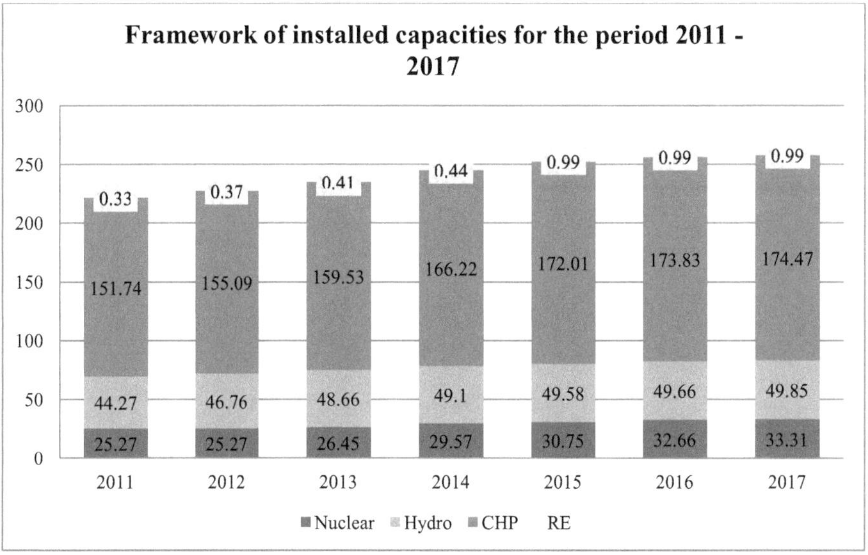

Figure 17: Framework of installed capacities for the period 2011-2017[22]

2.3 Renewable energy potential

In this chapter it is going to be discussed the potential of renewable energy in Russia. The economic potential, considered by different scientists, can be changed in the near future, if the Russian Government develops broader legislation to support the development of renewable energy.

Wind energy potential

Russia is one of the largest countries in the world and is situated in different climatic zones, which results in high wind energy potential. Mostly the highest wind energy potential is

---

[21] Scheme and program of IES of Russia development for the period 2011-2017, from 29.08.2011

[22] Scheme and program of IES of Russia development for the period 2011-2017, from 29.08.2011

concentrated along the Russian Federation seacoasts, in the vast territories of steppes and in the mountains. Regions most favourable for wind energy use include the North and Far East parts of Russia (see wind atlas in figure 18).

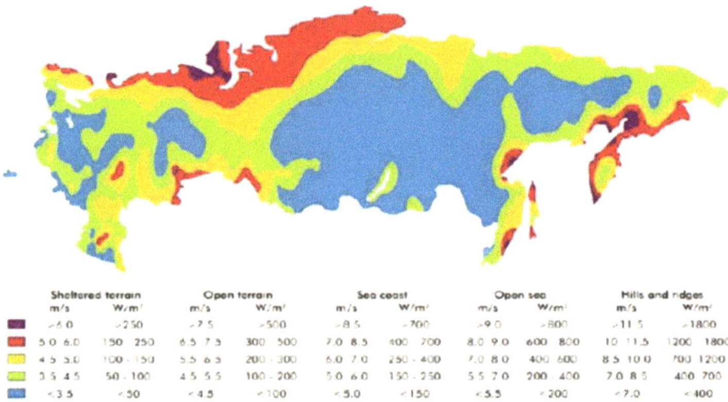

Figure 18: Russian Wind resources (Starkov A., Landberg L., Bezroukikh P.P.,Borisenko M.M., 2000)

The highest mean wind speeds are found along the coasts of the Barents and Kara seas, the Bering Sea and the Sea of Okhots. Other areas with relatively high wind speed (5-6 m/s) include the coasts of the East-Siberian, Chukchi and Laptev seas to the north and the Japan Sea to the east. Slightly lower wind speeds (3.5-5 m/s) are found on the coasts of the Black, Azov and Caspian seas in the south and on the White Sea in the north-west. Good resources are also found in the low and middle Volga regions, the Urals, the steppe areas of West Siberia, and around the Baikal Lake. The lowest mean wind speed occurs over East Siberia in the Lena-Kolyma core of the Asian anticyclone (Starkov A., Landberg L., Bezroukikh P.P.,Borisenko M.M., 2000).

The Russian Engineering Academy and Institution of Energy Strategy issued the "Handbook of Renewable Energy Resources of Russia", with a detailed consideration of each region, edited by Professor Bezroukikh P.P. The handbook considers the gross wind potential of the country to be 2606.6PWh, with a technical potential of 6516.61 TWh, and economic potential of 32.57 TWh (Bezroukikh, 2007, p. 85).

Wind energy can be exploited in many parts of Russia, including Arkhangelsk, Astrakhan, Volgograd, Kaliningrad, Magadan, Novosibirsk, Perm, Rostov, the Tiumen regions, Krasnodar, Khabarovsk, the maritime territories, Dagestan, Kalmykia and Karelia. The highest wind potential is on the area of the Far East, Ural and Siberia federal regions. (Bezroukikh, 2007, pp. 80-85).

The majority of the potential is found in the territories, where population density is less than one person per square kilometre. Thus, in many of the windiest places, wind energy can be used as an energy source for small isolated consumers.

Solar energy potential

Solar radiation depends largely on latitude, i.e. it is strongest at the equator and diminishes towards the poles. Russia is located between 41 and 82 degrees north latitude, and solar radiation levels vary considerably. According to Russian estimates, the average solar radiation in the remote northern areas is 810 kWh/m^2 per year, while in the southern regions it is more than 1400 kWh/m^2 per year. Solar radiation levels also exhibit great seasonal variations. For example, at latitudes of 55 degrees solar radiation is 1.69 kWh/m^2 per day in January, and 11.41 kWh/m^2 per day in July (Karabanov, 2001)

Figure 19 represents Solar Resources in Russia.

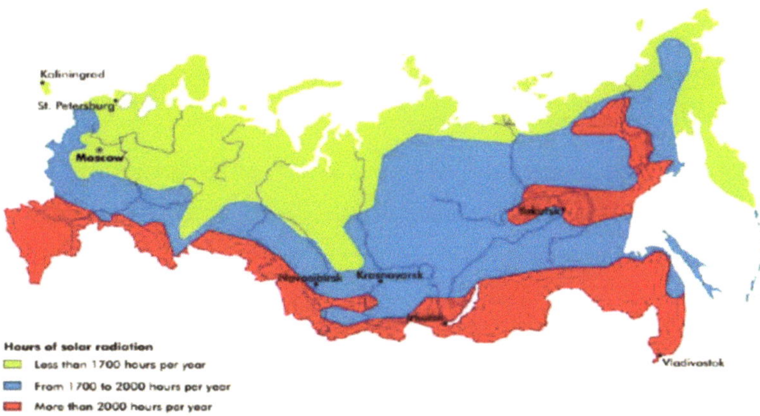

Figure 19: Solar Resources in Russia (Ecoclub, 2007)

Bezroukikh P.P. estimates the gross solar energy potential in Russia at 2205.4 Gtoe or 25648.81 PWh, technical potential at 9675.6 Mtoe or 112.5PWh and economic potential at 2.5 Mtoe or 29.3TWh (Bezroukikh, 2007, p. 73). The greatest solar energy potential is in the south-west (North Caucuses, the Black and Caspian Sea regions) and in Southern Siberia and the Far East. Regions with good solar resources include: Kalmykia, Stavropol, Rostov, Krasnodar, Volgograd, Astrakhan and other regions in the south-west, and Altay, Maritime, Chita, Buryatia and other regions in the south-east.

Biomass energy potential

Russia's biomass resources include vast forests, open woodlands, agricultural and wood waste. According to Bezroukikh P.P. the technical potential of available biomass includes: 1.46 Mtoe of poultry farming waste, 17.61 Mtoe of livestock farming waste, 55.23 Mtoe of crop production waste, 7.42 Mtoe of processing industry waste, 9.58 Mtoe of hard domestic waste, 0.71 Mtoe of sediment sewage, 48.44 Mtoe of lumbering and woodworking waste. In total it is 140.45 Mtoe of technical biomass potential in Russia or 1633.4 TWh. (Bezroukikh, 2007, p. 196)

In northwest Russia, residues from sawmills and the pulp and paper industry could supply as much as 45 to 50 TWh/y in the regions of Murmansk, Arkhangelsk, Kerelia, Vologda, Komi, Pskov, Novgorod and Leningrad. The Leningrad regional Forest Committee estimates that the total annual production of wood waste in Leningrad Region is 250,000 m^3 (12% of annual wood processing), of which one third to one half remains unused. Karelia, Vologda, Komi and Arkhangelsk each produce 2.5 to 7.5 million m^3 of wood waste annually (Bezroukikh, 2007, pp. 86-196). Figure 20 represents Wood Industry Location and Wood Flow Patterns of Russia.

Bezroukikh P.P. estimates the total gross biomass potential at 467.56 Mtoe or 5437.7 TWh, technical potential at 140.45 Mtoe or 1633.4 TWh, and economic potential at 69.07 Mtoe or 803.2 TWh (Bezroukikh, 2007, p. 197).

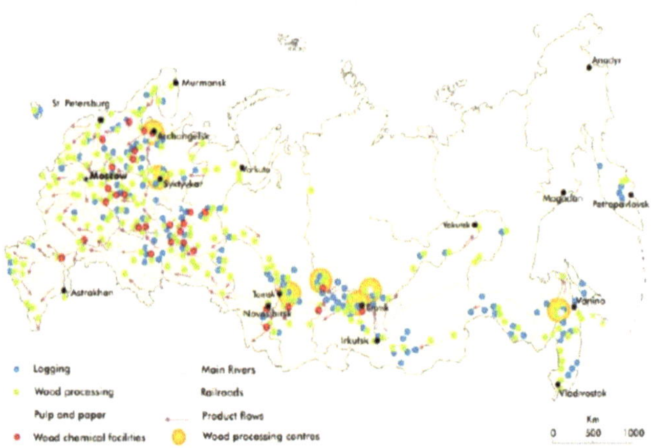

Figure 20: Wood Industry Location and Wood Flow Patterns of Russia (Nilsson, S. and Shvidenko, A., 1998)

Geothermal energy potential

Geothermal potential in Russia has a sparse pattern; the area of this potential includes the Kamchatka peninsula, the Kuril Islands, the Northern Caucasus, Dagestan, the Central region of Russia, the West Siberian plate, lake Baikal, the Krasnoyarsk territory, Chukotka, and Sakhalin. The highest geothermal potential is on the Kamchatka peninsula and the Kuril Islands. Kamchatka has 127 volcanoes, 22 of which are active, some 150 thermal spring groups and 11 high-temperature hydrothermal systems. The Kuril Islands have 100 volcanoes, including 21 active. The other areas of Russian geothermal potential have temperatures ranging from 50 to 200 C° at depths of 200 to 3000 metres. Some additional territories – the East-European and Siberian platforms, the belts of Ural, Altai and Saiany – contain geothermal potential, although, temperatures are 50-70 C° and they are significantly deeper - 3 kilometres depth (Povarov O.A., 2002). Figure 21 represents geothermal resources in Russia.

Energy production from technical potential of geothermal resources is 11868.7 Mtoe per year or 138 PWh, economic potential is estimated at a level of 114.9 Mtoe per year or 1336 TWh. Also, geothermal resources, prepared for industry development, are estimated at a level of 0.85 Mtoe per year or 9.89 TWh. (Bezroukikh, 2007, p. 218)

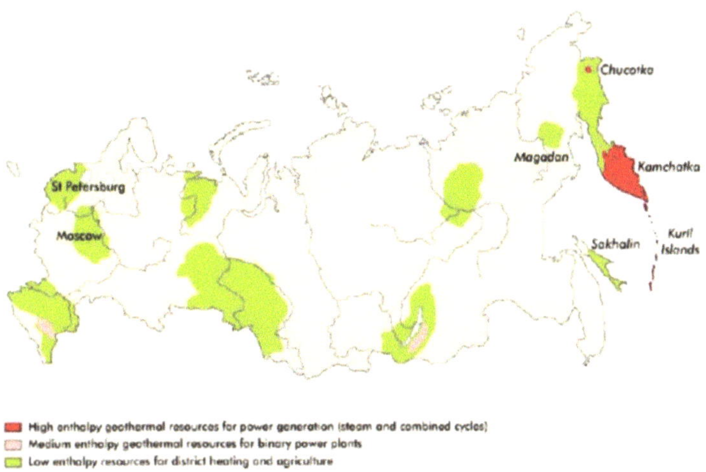

Figure 21: Geothermal resources in Russia (Povarov O.A., 2002)

Hydro energy potential

Russia has the second highest level of mean annual river runoff in the world, after Brazil. The number of rivers exceeds two million, and there are countless lakes and storage reservoirs. The highest potential is concentrated in the area of the Far East, Eastern Siberia, North Caucasus and the western part of the Urals. The European part of Russia has approximately 25% of overall potential. The average runoff varies according to water resources concentration; e.g., the runoff of the Northern Caucasus exceeds 2000 mm and the runoff of the European part of Russia ranges from 300 to 400 mm in the north-west to zero in the south-east. The dry period in the country mostly occurs between April and June (Martynov, A., Artukhov, V., Vinogradov, V., 1997). Figure 22 shows the hydro potential of different territories in Russia.

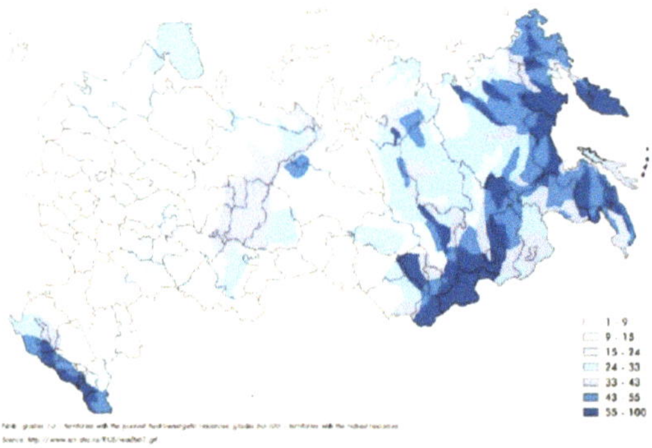

Figure 22: Hydro resources in Russia (Martynov, A., Artukhov, V., Vinogradov, V., 1997)

Bezroukikh P.P. estimates total gross hydro potential at 401.6 Mtoe or 1280.32 TWh, technical potential at 126.5 Mtoe or 371.11 TWh, and economic potential at 69.7 Mtoe or 205.09 TWh. (Bezroukikh, 2007, p. 78)

The more detailed figures of each kind of RE sources potential in every federal region of the country can be found in Annex 1, according to the Handbook of Renewable Energy Resources of Russia.

2.4 Potential of energy efficiency measures in electricity generation, transmission and consumption

In this chapter the potential of energy efficiency measures in the electricity generation, transmission and consumption will be described according to the government program "Energy savings and energy efficiency improvement for the period till 2020". This program was chosen as the main source of energy efficiency potential, as all governmental companies developed their energy efficiency and innovative programs according to this Government program.

Potential measures in electricity generation

Potential of energy efficiency in electricity generation overall is considered in electricity generation by CHP power plants. It can be explained by the high share of CHP power plants, making up more than 65 % of total electricity generation in the sector (Federation, 2012).

Thus, it is considered that energy efficiency measures on the gas PP will be through the decommissioning of worn-out gas power plants; construction of gas PP, which use gas-turbines and steam technologies; modernization of existing power plants which use modern innovative equipment with efficiency not less than 55% up to 2015 and not less than 60% up to 2020 (Government, 2010, pp. Annex 6, 20-22).

These measures will allow savings of 148.76 Mtoe or 1730 TWh and 244.56 m ton of CO_2 during the period of 2011-2020. Also planned is the provision of reconstruction and modernization in 95 power plants; 8 power plants every year up to 2015 and 11 power plants every year from 2015 onwards up to 2020 (Government, 2010, pp. Annex 1,pp. 1-2; Annex 4,p. 1; Annex 6, p. 20; Annex 7, pp. 1-2).

The next measure considered is the implementation of coal PP, through the decommissioning of worn-out coal PPs; construction and modernization of PPs with usage of new innovative technologies, such as super high steam parameters, boilers with circulating fluidized bed, etc., with efficiency of not less than 43% up to 2015 and not less than 48% up to 2020. It will save 114.68 Mtoe or 1334 TWh and 319.96 m ton of CO_2 in the period 2011-2020. To achieve these figures it is necessary to modernize 3 PPs each year, in total 30 Coal PPs (Government, 2010, Annex 1, pp. 1-2; Annex 4, p. 1; Annex 6, p. 20; Annex 7, pp. 1-2).

Energy efficiency in diesel PP will also be implemented through the decommissioning of worn-out PPs and construction of new diesel PPs which use innovative technologies. Modernization of existing PPs has to be done with the reduction of fuel consumption up to 356 g of oil equivalent per kWh. These measures will result in savings of 5.09 Mtoe or 59.2 TWh and 12.61 m ton CO_2. These targets will be achieved through the reconstruction and modernization of 100 PPs during the period 2011-2020 (Government, 2010, Annex 1, pp. 1-2; Annex 4, p. 2; Annex 6, p. 20; Annex 7, pp. 1-2).

The overall energy efficiency measures with the planned results are represented in table 5.

Table 5: Expected results of energy efficiency measures in electricity generation[23]

Type of PP	Quantity to modernize	Savings TWh	m ton of CO_2
Gas PP	95	1730	244.56
Coal PP	30	1334	319.96
Diesel PP	100	59.2	12.61
Total PP	**225**	**3123**	**577.13**

Potential measures in electricity transmission

The main measures in electricity transmission are expected to reduce losses of up to 8-9% by 2020 and to improve systems of commercial and technical metering of transmission and consumption (Government, 2010, p. 12). The energy efficiency measures in electricity transmission have a wide range. These include transmission lines of high, middle and low voltage classes, cables, transformers, sub-stations and technical equipment of electricity consumption.

Table 6 represents a summary of planned measures in the transmission system of the country.

The measures described in table 6 will allow a saving of 44.28 Mtoe or 515 TWh and 93.71 m ton CO_2 (Government, 2010, Annex 7, p. 2; Annex 4, p. 2).

[23] Derived by the author from Government program "Energy savings and improvement of energy efficiency for the period up to 2020"

Table 6: Planned energy efficiency measures in electricity transmission (Government, 2010, pp. Annex 6, pp. 20-22)

Name	unit	Quantity for the period 2011-2020
Reconstruction of transformer sub station	pcs.	142667
To exchange transformers	pcs.	288824
Reconstruction of over headlines:		
110 kV and above	Km	31840
35-6 kV	Km	589021
0.38 kV	Km	347506
Reconstruction of cables:		
35-6 kV	Km	48778
0.38 kV	Km	26012
Reconstruction and/or expansion of sub-stations	pcs.	540
Implementation of AIMSCEPA[24] by consumers	pcs.	7004364
Replacement of single-phase induction electric meters	pcs.	541966

Potential measures in electricity consumption

Consumption measures are considered in the industrial and residential sector.

The energy efficiency measures in the industrial sector are mostly considered to come from the integration of more efficient technologies in the sector, such as (Government, 2010, pp. Annex 6, pp. 27-29):

- Quantity of engines with high efficiency class: from 1.92 million pcs in 2011 to 6.96 million pcs in 2020
- Quantity of engines with frequency-controlled drive: from 0.21 million pcs in 2011 to 0.59 million pcs in 2020
- Productivity of compressed air systems, with characteristics equal to the best world practice: from 7110 million m^3 per year in 2011, up to 35,666 million m^3 per year in 2020
- Share of efficiency lamps in industrial lighting: from 37% in 2011 up to 73% in 2020
- Share of energy efficiency systems in the industrial lighting: from 19.9% in 2011 till 49.6% in 2020

These measures will allow a saving of 62.7 Mtoe and 127.92 m ton CO_2 for the period 2011-2020. Also, realization of energy efficiency measures in the manufacturing sector will allow

[24] AIMSCEPA – Automatic information and measuring system of commercial electric power accounting is set of hardware and software which provide remote data collection, storage and processing of data in the grid.

a saving of 70.63 Mtoe and 87.93 m ton CO_2 for the period 2011-2020 (Government, 2010, Annex 4, pp. 7-9; Annex 7, pp. 8,9).

The energy efficiency measures in the residential sector mostly consist of heating measures and electricity saving measures. Thus, it is necessary to construct 13 million m^2 of new municipal buildings with high parameters of heat insulation; to provide major repairs in municipal buildings in 192.3 million m^2, share of well insulated buildings has to increase from 12% in 2011 up to 88% in 2020; installation of energy efficient lamps, 1481 million lamps for the period 2011-2020; exchange 17.7 thousand units of worn-out gas boilers. These measures will save 115.95 Mtoe and 231.74 m ton CO_2 for the period 2011-2020 (Government, 2010, pp. Annex 6, pp. 29,30; Annex 4, pp. 11, 12; Annex 7, pp. 11, 12).

Energy efficiency measures in the residential sector include installation of different metering systems of heat (2433.8 thousand units), hot water (24836.3 thousand units) and gas (10960 thousand units); major repair of apartment buildings (340 million m^2); exchange of worn out fridges and refrigerators (140 million units); exchange of worn out washing machines (117 million units); exchange of worn out gas boilers (6.2 million units). The share of energy efficient lighting has to increase from 15% in 2011 up to 83% in 2020. These measures will allow savings of 97.83 Mtoe and 190.04 m ton CO_2 (Government, 2010, pp. Annex 6, pp. 31, 32; Annex 4, pp.12-14; Annex 7, pp. 12-14).

3 THE MARKET POTENTIAL FOR RENEWABLE ENERGY AND ENERGY EFFICIENCY

This chapter provides a consideration of the main players in the market promoting renewable energy and energy efficiency in the sector and areas of the market for further renewable energy and energy efficiency development.

3.1 Main players in the market promoting renewable energy and energy efficiency

The main players in the market, during the research and analysis, were divided into government and private enterprises and international financial institutions investing in the renewable energy and energy efficiency part of the market.

Government and private enterprises

Potential investors, who are interested in investing in RE technologies in the Russian market, can establish stable cooperation with the five big government companies of the Russian Energy Sector. These companies represent all parts of the sector: generation, transmission and distribution.

The generation part is represented by open joint stock companies "Rushydro", "Energy systems of the East" JSC (ESE) and "Inter RAO UES" (Inter RAO).

Transmission and distribution are represented by "Federal grid company of Integrated Energy Systems" JSC (FGC) on the level of high voltage transmission systems and by "Holding of inter-regional distribution grid companies" (IDGC) on the level of transmission systems up to 220 kV.

The transmission and distribution parts are also represented by the ESE and Inter RAO.

Since ESE is a vertically integrated company, which combines generation, transmission and distribution and Inter RAO is the owner of generation and distribution facilities after a restructuring program.

"Rushydro" JSC (Rushydro)

Rushydro is one of the big companies in the energy sector, which owns only hydro power plants with total installed capacity of 35.2 GW and production of 77008 GWh in 2011. The main owner of the company is the Russian Federation – 60.38% and the rest of the assets are distributed within private companies (Rushydro, 2012). Figure 23 represents distribution of the Rushydro share:

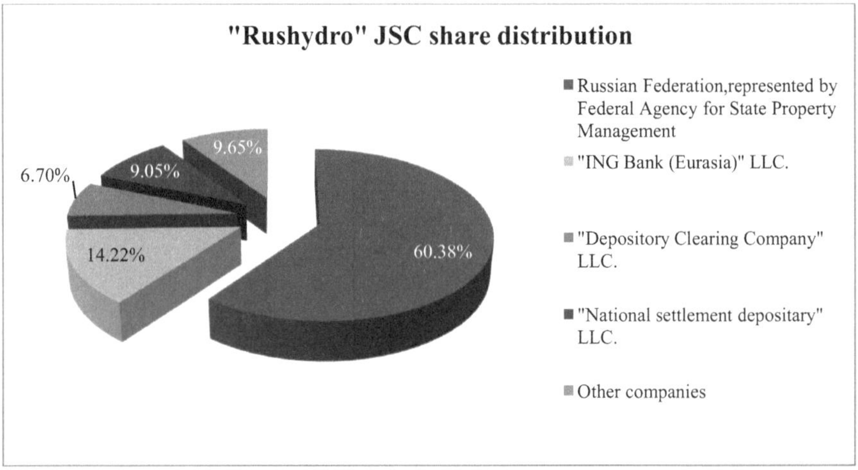

Figure 23: "Rushydro" JSC share distribution[25].

Rushydro has 20 branches in different regions of the country. Table 7 represents each branch with quantity of power plants in the branch, location in the integrated energy system and installed capacities:

Table 7: Branches of "Rushydro" JSC[26]

№	Name	IES	Quantity	Installed Capacity, MW
1	**Bureiskaya HPP**	**Far East IES**	**1**	**2010**
2	**Volga HPP**	**Volga IES**	**1**	**2592.5**
3	**Votkinskaya HPP**	**Ural IES**	**1**	**1020**
4	**Zhigulevskaya HPP**	**Volga IES**	**1**	**2341**
5	**ZagorskayaPumped Storage PP**	**Centre IES**	**1**	**1200**
6	**Zeiskaya HPP**	**Far East IES**	**1**	**1330**
7	**Nizhegorodskaya HPP**	**Centre IES**	**1**	**520**
8	**Novosibirsk HPP**	**Siberia IES**	**1**	**455**

[25] Derived by the author, based on information from http://www.rushydro.ru/investors/capital/structure

[26] Derived by the author, based on information from http://www.rushydro.ru/company/structure

№	Name	IES	Quantity	Installed Capacity, MW
9	**Saratov HPP**	**Volga IES**	**1**	**1360**
10	**Sayano-Shushenskaya HPP**	**Siberia IES**	**2**	**6721**
11	**Kamskaya HPP**	**Siberia IES**	**1**	**522**
12	**Karachaevo-Cherkeskiibranch**	**South IES**	**1**	**160**
13	**Cascade of Verchne-Volzhskih HPP**	**Volga IES**	**2**	**456.6**
14	**Cheboksarskaya HPP**	**Volga IES**	**1**	**1370**
15	**Irganaiskaya HPP**	**South IES**	**1**	**400**
16	**Dagestan Branch:**	**South IES**	**12**	**1781.9**
	Chirkeskaya HPP			1000
	Irganaiskaya HPP			400
	Miatlinskaya HPP			220
	Chiryurtskaya HPP-1			72
	Chiryurtskaya HPP-2			9
	Gelbachskaya HPP			44
	Gergebilskaya HPP			17.8
	Gunibskaya HPP			15
	Kurushskaya SHPP			0.5
	Achtynskaya SHPP			1.8
	Agulskaya SHPP			0.6
	Maginskaya SHPP			1.2
17	**Kabardino-Balkarbranch:**	**South IES**	**6**	**155.4**
	Baksanskaya HPP			25
	Cascade of Nizhne-Cherkesskih HPP			125
	Akbashskaya SHPP			1
	SHPP-3			3.5
	Muholskaya SHPP			0.9
18	**Cascade of Kubanskih HPP:**	**South IES**	**10**	**476.6**
	Group of Kurshavskih HPP			
	Pumpedstorage HPP			15.9
	HPP-1			37
	HPP-2			184
	Group of Barsuchkovskih HPP:			
	HPP-3			87
	HPP-4			78
	Svistuhinskaya HPP			11.8
	Group of Sengileevskih HPP:			
	Sengileevskaya HPP			15
	Egorlykskaya HPP-1			30
	Egorlykskaya HPP-2			14.2
	Novotroizkaya HPP			3.7
19	**Severo-Osetinskiibranch:**	**South IES**	**5**	**76.9**
	Ezminskaya HPP			45
	Gizeldonskaya HPP			22.94
	Dzaudzhikauskaya HPP			9.2
	Bekanskaya HPP			0.504
	Kora-Ursdonskaya			0.63
20	**Corporative University of Hydro Energy**			

Rushydro is obliged by the Government to promote renewable energy development. The company considers a broad band of options by type and capacities of RE design and installation.

Thus, based on the Rushydro strategy up to 2015 (with a possible extension up to 2020) it has to install 10 GW of new conventional HPP, approximately 140 MW of RE PP, which include HPP, geothermal PP and wind PP and to design 500 MW of small HPP and 1000 MW of wind PP. The company also considers the possibility of installing more RE PP, in case of further government support, and approval of statements, which were explained in the Chapter *Current and future electricity supply* (Rushydro, 2012).

„Federal grid company of Integrated Energy Systems" JSC (FGC)

FGC is a company providing transmission services to the integrated national electricity system. The FGC is a natural monopoly company and has 124500 km of transmission lines and 311000 MVA of transformer capacities (FGC, 2012). The major share of the company belongs to the government (see figure 24).

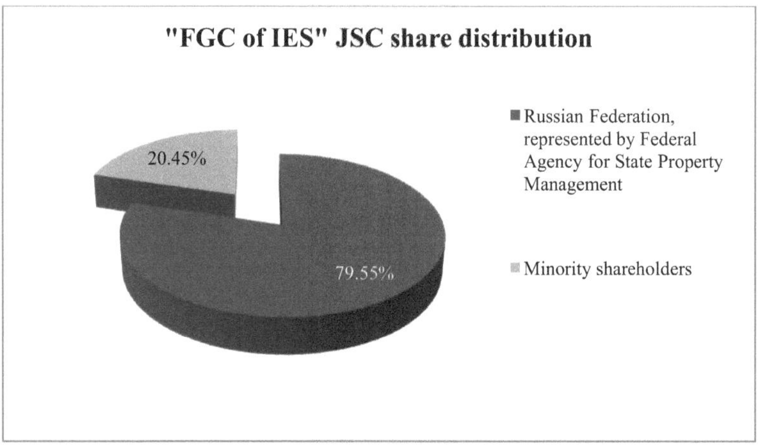

Figure 24: "FGC of IES" JSC share distribution[27].

The company has 8 branches of transmission lines in all regions of the country (FGC, 2012):

- Central Power Grids (PG): includes 9 transmission enterprises
- North-West PG: includes 6 transmission enterprises

[27] Derived by the author, based on information from www.fsk-ees.ru

- Volga PG: includes 4 transmission enterprises
- South PG: includes 5 transmission enterprises
- Ural PG: includes 4 transmission enterprises
- North Siberia PG: includes 4 transmission enterprises
- Siberia PG: includes 7 transmission enterprises
- East PG: includes 3 transmission enterprises

It has 22 subsidiaries and affiliated companies, with participation share from 33.33% to 100%. These companies include service companies, a scientific institute, an educational institute, a technological platform and others (FGC, 2012).

Far East energy company "Energy systems of the East" JSC (ESE)

ESE is the biggest energy company in the Far East, with a government share of more than 60%, but direct government share is only 3.28% by Federal Agency for State Property Management; the main share has Rushydro, one of the largest governmental energy enterprises – 65.75% (see figure 25).

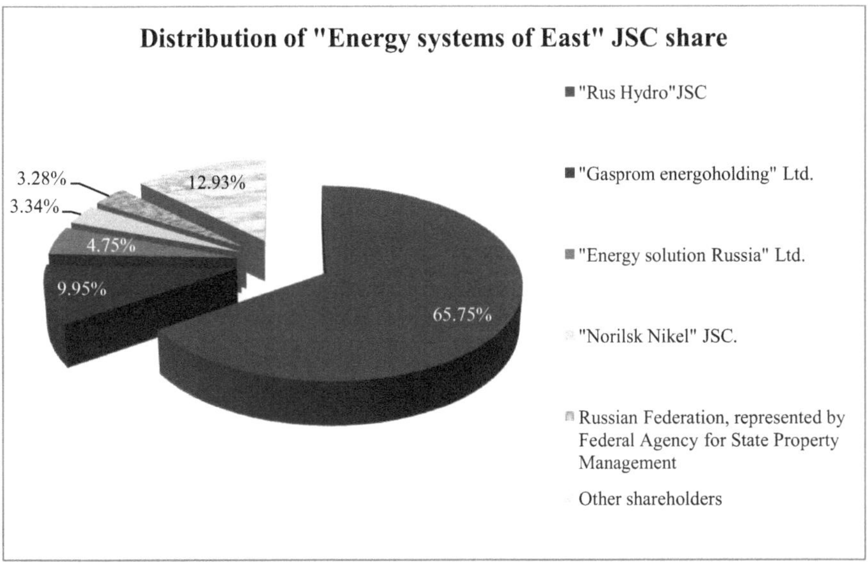

Figure 25: Distribution of "Energy systems of East" JSC share[28].

This is one of two energy companies in the Russian energy sector which combines vertically integrated energy companies: generation, transmission and distribution. The second

[28] Derived by the author, based on information from www.rao-esv.ru

company is “Inter RAO UES”. The ESE includes 8 territorial energy systems divided into the integrated energy systems of the East (IES East) and isolated systems ("Energy systems of East" JSC, 2012):

1. Amurskaya
2. Primorskaya
3. Khabarovskaya
4. Camchatskaya
5. Magadanskaya
6. Chukotskaya
7. Sahalinskaya
8. Yakutskaya

Amurskaya, Primorskaya, Khabarovskaya and the southern part of Yakutskaya energy systems are integrated into IES East and operate in parallel conditions. The rest of the systems function in isolation and operate in individual conditions ("Energy systems of East" JSC, 2012).

92% of installed capacities of the ESE is represented by CHP and 8% is by hydro PP.
The share of worn out equipment from overall assets of ESE is 60-70%, including 40% of equipment older than 40 years. Although around 30% is still efficient and overall equipment is on a high level of obsolescence (ESE, 2009).

“Inter RAO UES” CJSC (Inter RAO)

Inter RAO is a diversified power holding involved in electricity and heat generation and supply, export-import of electricity and engineering. Inter RAO operates and manages 27 CHP, 2 HPP and 1 wind farm of 30 MW. The total installed capacity of the company was 29 GW with electricity output of 117 TWh in 2011. Inter RAO is also a leading Russian export-import operator accounting for 97% of Russian electricity export in 2011. The company operates in over 14 countries, including Belarus, Lithuania, Georgia, Armenia, Kazakhstan, Tajikistan, Moldova and Finland. It owns 9 distribution companies that supply 314000 legal entities and 9.8 million private consumers (UES, 2012). Figure 26 represents the broad geographical coverage of the company:

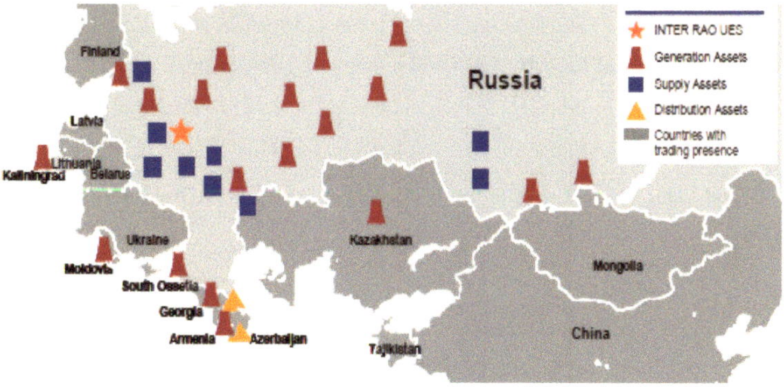

Figure 26: Operation area of "Inter RAO UES" CJSC (UES, 2012).

The Company's shareholders are mostly government companies. FGC, Federal Agency for State property Management and Rosenergoatom have the highest participation (see figure 27).

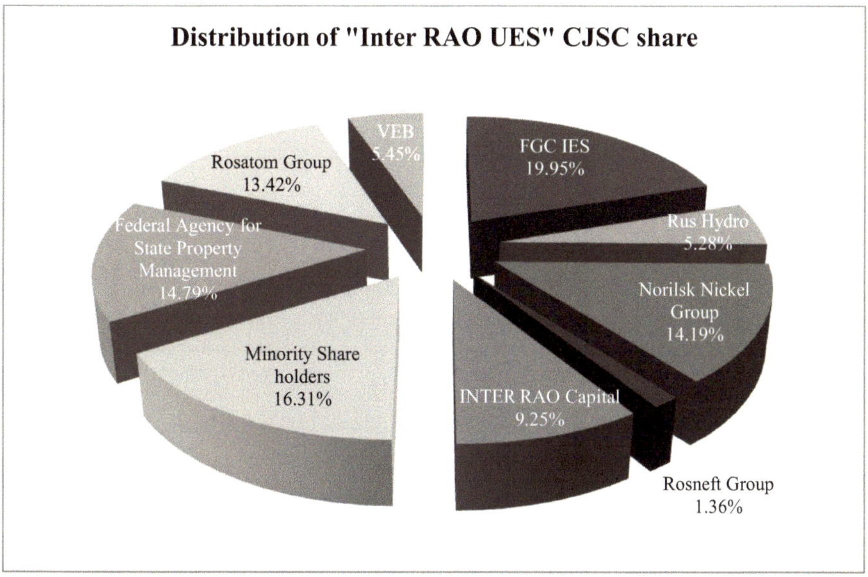

Figure 27: Distribution of "Inter RAO UES" CJSC share (UES, 2012).

Inter RAO, itself, has a major share in three big companies: two generation companies of the wholesale market – "OGK-1" JSC (75%) and "OGK-3" JSC (82%) and 1 territorial generation company "TGK-11" JSC (68%). The company plans to increase its share of OGK-1 and OGK-3 up to 100% before 2013 (UES, 2012).

"Holding of inter-regional distribution Grid Company" JSC (IDGC)

IDGC is a company that comprises interregional and regional distribution grid companies (IDGC/RDGC), research and development institutes, design and construction institutes, and construction and sales entities. Ninety-seven branches of IDGC/RDGC are situated in 69 regions of the Russian Federation (IDGC, 2012).

IDGC has share in 44 joint-stock associated and affiliated companies, including 11 IDGC and 5 RDGC. The controlling shareholder of IDGC Holding is the Government - 53.69% (2011) (see figure 28).

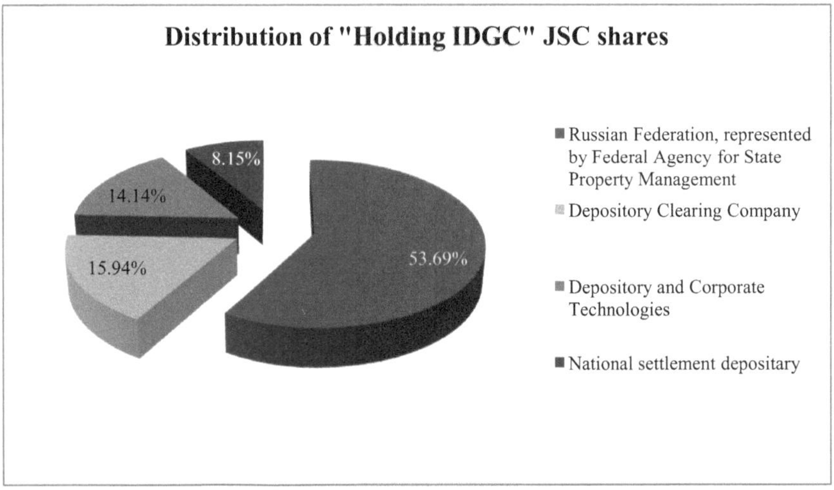

Figure 28: Distribution of "Holding IDGC" JSC shares[29].

The operational area of IDGC companies covers a power network of ten voltage types, ranging from 0.4 kV to 220 kV. The total spread of networks of IDGC operating subsidiaries exceeds 2.1 million km. IDGC ranks among the world's largest electrical grid companies in terms of network spread and customers (IDGC, 2012).

[29] Derived by author, based on information from http://www.holding-mrsk.ru/investors/capital/share/

Technological platforms (TP)

The Russian Ministry of Energy, Rushydro, FGC and other entities of the market additionally promote renewable energy and energy efficiency through Technological platforms. Technological platforms are the area of the market which provides scientific, advisory and financial support for further development of RE and energy efficiency in the Russian energy sector.

There are different technological platforms, such as: “Small distribution energy”, “Smart Russian energy system”, “Renewable energy advanced technologies”[30] “Smart Russian energy system” TP has been established with the support of and headed by the Ministry of Energy. Among other services (advisory, scientific, etc.) the platform supports RE and energy efficiency projects implementation, also with partial investment from international organizations (Agency, 2012).

Private enterprises

The Russian energy sector has six wholesale generation companies and 14 territorial generation companies, which are different from the government enterprises. Most of these companies are focused on energy production from heat PPs, some of them have in their assets big hydro PPs, and one territorial generation company, TGC-8, which owns only big hydro PPs.

During the study of investment programs of these companies, projects promoting RE technologies could not be found. But half of the listed companies reconstructed their heat PPs and are planning to construct new PPs with installation of new energy efficiency technologies. A list of private generation companies with heat PPs in their assets can be found in Annex 2.

---

[30] www.e-apbe.ru, www.rosenergo.gov.ru,
http://www.rushydro.ru/company/energy_efficiency/emerging_technologies

Investment in the Russian energy sector by the international financial institutions

This sub-chapter represents 4 international financial institutions, implementing projects in the Energy sector of Russia.

European Bank for Reconstruction and Development (EBRD).

The EBRD financed different projects in the Russian energy sector from 2005 to 2011 in the private and public area. During the research only projects implemented in the public area were investigated. Nowadays there are no projects under consideration; the last signed project in the energy sector was on 21st October 2011. The active projects are performing in the sector of "Power and Energy" and "Energy Efficiency".

Projects in the sector of "Power and Energy" (PE) are implemented for reconstruction and modernization of HPPs, which belong to Rushydro; for restructuring of ESE, where Rushydro has the majority of the shares; and for modernization of substations that belong to FGC.

The "Energy Efficiency" (EE) projects are mostly implemented in the area of modernization of district heating technologies in the Far East region. Table 8 represents EBRD public projects under implementation.

Table 8: Energy projects financed by EBRD[31].

Date	*Financed company*	*Description*	*Sector*	*Project Cost*	*Amount*
21 Oct. 11	Rushydro	Balance sheet restructuring of ESE. The EBRD will engage with RusHydro on restructuring of the Far East energy system and facilitate completion of the power sector reform.[32]	PE	RUB 20 bn	Senior loan of RUB 8 bn. (equivalent to ~ EUR 179.4 m)
7 Jul. 11	FGC	EBRD participating in a bond emission, in order to contribute to the modernization of Kamala substation located in Siberia.[33]	PE	2011 bond issuing program - RUB 125 bn	Purchase up to RUB 4.25 bn subject to market conditions (equivalent to ~ EUR 95.3m)
1 Nov. 10	FGC	EBRD participating in a bond issue, in order to contribute to the	PE	2010 bond issuing program - RUB 50 bn	Purchase up to RUB 5 bn subject to market conditions

[31] Derived by author, based on information from www.ebrd.com

[32] http://www.ebrd.com/english/pages/project/psd/2011/43190.shtml

[33] http://www.ebrd.com/english/pages/project/psd/2011/42996.shtml

Date	Financed company	Description	Sector	Project Cost	Amount
		modernization of Pakhra substation located in the Moscow region.[34]			(equivalent to ~ EUR 112.1m)
22 Oct. 10	Rushydro	EBRD participating in a bond issue, in order to contribute to rehabilitation project of Volzhskaya HPP.[35]	PE	Bond issuing program - RUB 20 bn	Purchase in the amount of RUB 4.5 bn (equivalent to ~ EUR 100.9 m)
27 Feb. 09	"YuTEC" (Yugra Heat and Energy Company), the power transmission and distribution company, owned by the Khanty-Mansi Autonomous Okrug (Okrug).	Rehabilitation and upgrade of the municipal local heating systems through refurbishment of up to 164 small electricity and heat generating facilities (i.e. diesel fired generators and boilers) and rehabilitation of adjacent distribution networks in small settlements of the Okrug. The regulatory framework improvement component for the municipal and housing services sector of Okrug.[36]	EE	RUB 2,000 m (equivalent to ~ EUR 44 m).	Loan up to RUB 1,500 m (equivalent to ~ EUR 33.6 m)
24 Apr. 07	the municipal services management company "UGH"	The proposed project is a loan to UGH to finance priority investment program to improve efficiency of the municipal discrict heating and water services in the city of Pyt-Yakh of the Khanty-Mansi Autonomous Okrug.[37]	EE	RUB 550 m (equivalent to EUR 16 m)	The loan to UGH in the amount of RUB 350 m (equivalent to EUR 10.2 m).
13 Mar. 07	Russian Federation with proceedings to ME Kaliningradt eploset	Rehabilitation of the district heating network in the City of Kaliningrad.[38]	EE	EUR 20.8 m	Sovereign loan of USD 12 m (EUR 11.2 m). To be co-financed with a grant from the Northern Dimension Environmental Partnership (NDEP) of EUR 7.3 m
29 Mar. 06	The municipality of Surgutsky Rayon	Improvement of municipal district heating, water and wastewater infrastructure facilities and services in small settlements of Surgutsky Rayon.[39]	EE	RUB 1,400 m (equivalent to ~ EUR 40.5 m)	The loan up to RUB 810 m (equivalent to ~ EUR 23 m)

[34] http://www.ebrd.com/english/pages/project/psd/2010/41537.shtml

[35] http://www.ebrd.com/english/pages/project/psd/2010/41886.shtml

[36] http://www.ebrd.com/english/pages/project/psd/2009/38604.shtml

[37] http://www.ebrd.com/english/pages/project/psd/2007/37767.shtml

[38] http://www.ebrd.com/english/pages/project/psd/2007/37966.shtml

[39] http://www.ebrd.com/english/pages/project/psd/2006/36340.shtml

Date	Financed company	Description	Sector	Project Cost	Amount
16 Mar. 06	Rushydro	Financing of the refurbishment of several units of the 10,000 MW Volga-Kama hydro cascade and development of renewable sources of energy.[40]	PE	RUB 11.4 bn	Senior Debt Facility of up to RUB 6.3 bn (190 m Euro equivalent)
4 Aug. 05	the City of Yakutsk, State Unitary Enterprise "Yakutsk Vodokanal" and State Unitary Enterprise of Communal and Housing Services (GUP ZHKH)	To reduce operating costs and finance general rehabilitation and upgrading of the heating and water systems in the Yakutsk City.[41]	EE	RUB 2,500 m (EUR 71 m)	1. A loan up to RUB 300 m (EUR 8.5 m) to the City; 2. A loan up to RUB 350 m (EUR 10 m) to Vodokanal; 3. A loan up to RUB 1000 m (EUR 28.4 m) to GUP ZHKH
11 Apr. 05	the Ufa Municipal Company for Housing Management and Engineering Support	To finance the rehabilitation and modernization of existing Central Heating Sub-stations and the introduction of new, compact Individual Heating Sub-stations in residential apartment buildings in the City of Ufa.[42]	EE	Equivalent to €15 m.	Senior loan of €10 m (RUB 360 m)

The World Bank (WB)

By the data represented in the official sources of the WB, there is only one active project, where the energy sector has 52% of investment, the rest of investment goes towards the general water, sanitation and flood protection sector and sub-national government administration. There are also three proposed projects for further energy sector development. Two proposed projects are direct investments in the development of the energy sector and energy efficiency measures and one project is a loan to the State Corporation "Bank for Development and Foreign Economic Affairs (Vnesheconombank)" (VEB). The active and proposed short description of the projects can be found in table 9.

---

[40] http://www.ebrd.com/english/pages/project/psd/2006/36725.shtml

[41] http://www.ebrd.com/english/pages/project/psd/2005/35294.shtml

[42] http://www.ebrd.com/english/pages/project/psd/2005/35733.shtml

Table 9: Active and proposed energy projects financed by World Bank[43].

Starting Date	*Closing Date*	*Status*	*Financed company*	*Description*	*Sector*	*Project Cost*	*Amount*
26 Feb. 08	1 Jun14	Active	Federal Agency for construction and Housing and Communal service of RF (Rosstroy)	The 52% of the project covers energy efficiency in power sector. The energy efficiency concerns mostly district heating and hot water supply. [44]	Energy Efficiency in power sector	USD 206 m	USD 200 m
N/A	N/A	Proposed	VEB	The project is intended to finance transportation sector (40%), water, sanitation and food protection (30%) and energy sector (30%). Project objectives are to increase VEB capacity to identify, appraise, monitor and evaluate infrastructure investment and public-private partnership and to enable VEB to provide long-term financing to infrastructure sub-projects and public-private partnership. [45]	General energy sector	USD 1000 m	USD 500 m
N/A	N/A	Proposed	Gazprombank and Ministry of Energy/Russian Energy agency	Climate change (50%), Infrastructure services for private sector development (25%), Micro, Small and Medium enterprise support (25%). [46]	Energy Efficiency in power sector	USD 322.73 m	USD 322.73 m
N/A	N/A	Proposed	Gazprombank and Ministry of Energy/Russian Energy agency	Climate change (50%), Infrastructure services for private sector development (25%), Micro, Small and Medium enterprise support (25%). [47]	Energy Efficiency in power sector	USD 775 m	USD 300 m

[43] Derived by author, based on information from www.worldbank.org

[44] http://www.worldbank.org/projects/P079032/housing-communal-services-project?lang=en

[45] http://www.worldbank.org/projects/P125905/russian-development-bank-infrastructure-finance?lang=en

[46] http://www.worldbank.org/projects/P123692/russia-energy-efficiency-gef?lang=en

[47] http://www.worldbank.org/projects/P122492/russia-energy-efficiency-financing-project?lang=en

The European Investment Bank (EIB)

The EIB has two projects under implementation, related to the Russian energy sector. The first project was started in March 2010 for modernization of OGK-5 JSC. The second project was signed in December 2011. This project includes three parts, where one of them is energy efficiency measures, reducing the use of natural gas in the Mondi Syktyvkar pulp and paper mill. There are no listed projects under consideration or approval. Table 10 represents a short description of the two active projects:

Table 10: Energy projects financed by EIB[48].

Starting Date	*Status*	*Financed company*	*Description*	*Sector*	*Project Cost*	*Amount*
26 Mar10	Active	"OGK-5" JSC. (Enel), private sector	Upgrade of the Nevinnomyssk GRES power plant in Southern Russia, including the rehabilitation of existing generating capacity and building of a modern, efficient Combined Cycle Gas Turbine (CCGT) unit to replace two old gas-fired steam units.[49]	Energy Efficiency	EUR 500 m	EUR 250 m
20 Dec11	Active	Mondiplc	The project comprises the modernization and expansion of the existing integrated Mondi Syktyvkar pulp and paper mill located close to the city of Syktyvkar, Komi Republic, in the north-west of the Russian Federation. This includes pulp, paper and paper-board production, power generation and extension of the wood yard.[50]	Agriculture, Industry, Energy	EUR 100 m	Energy part EUR 25 m

The French Government Environmental Fund (FGEF) (Agence Francaise de Development)

The FGEF has implemented one project in the area of energy efficiency and renewable energy promotion in Eastern Europe and Russia. The name of the project is «Renforcement de capacités et appui à la miseen place d'un fonds dédié à l'efficacitéénergétique en Europe de l'Est» (Energy efficiency support and promotion in Eastern Europe).

---

[48] Derived by the author, based on information from www.eib.org

[49] http://www.eib.org/projects/pipeline/2008/20080333.htm

[50] http://www.eib.org/projects/pipeline/2010/20100198.htm

The project was part of the "Energy Efficiency 21" program. It was implemented to promote energy efficiency in Russia and several Eastern Europe countries (Ukraine, Belarus, etc.). The goal of the project was to raise awareness among actors and decision makers, to create a network of experts, to strengthen energy efficiency policies and to develop renewable energies. Lastly, it supported the creation of a 300 million euro public-private investment fund which makes it possible to provide incentives to the energy efficiency and renewable energy market in the region. The project was planned for completion in 2008 (Fund, 2004) Information of the project completion could not be found on the official sources of the Fund.

3.2 Areas of the market for renewable energy and energy efficiency development

This chapter considers potential in the sectors of renewable energies and energy efficiency.

Potential in the sector of renewable energies

The 2030 target of 4.5% renewable energy installed capacities in the electricity sector can be achieved by the companies already in 2017. Moreover, the sector has an opportunity to have in total 44.98 GW of new capacities in 2017, instead of 24 GW targeted in 2030. This capability has been found by comparing the RE economic potential in the electricity sector and the plans of the generation companies to install 9.9 GW of nuclear PPs and 34.4 GW of CHP up to 2017.

The Russian electricity sector is divided into seven integrated energy systems (IES), which are identical with the seven federal regions: North-West, Central, Volga, South, Ural, Siberia and the Far-East federal regions. The calculated capability of RE installations up to 2017 was made according to these 7 IES. Such distribution allows for estimation of the economic potential of RE technologies in each region and permits each region to generate its own electricity, to avoid additional costs for electricity transmission and distribution.

The estimated new capacities of Russia's IES were found in the "Scheme and program of IES of Russia development for the period 2011-2017". Overall, new capacities are planned for commissioning by 2017 and comprise of 50.05 GW, where 9.9 GW are nuclear PPs and 34.4 GW are heat PPs. The rest of the plants are hydro, wind, tidal and small hydro power plants. (SO, 2011, pp. 35, 207) (See Figure 29).

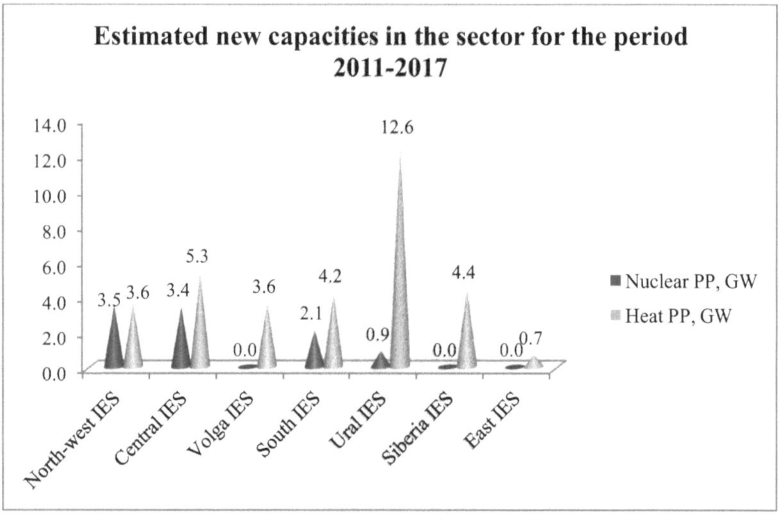

Figure 29: Estimated new capacities in the sector for the period 2011-2017[51]

In order to calculate the capabilities to install 44.3 GW of RE PPs instead of nuclear and heat power plants, the economic potential of RE resources in each federal region of the country was considered (See Table 11).

Table 11: Economic potential of RE resources in Russian federal regions[52]**.**

Name of Federal region	Economic potential, TWh						
	Solar	Hydro	Wind	Biomass	Biogas		
					Agric.	Landfill	Sewage
Central	0.35	1.57	0.36	28.6	80.01	27.56	2.44
North-West	0.18	8.89	2.16	57.22	8.6	10.47	0.93
South	0.14	10.3	0.88	32.1	143.9	11.98	1.05
Volga	0.43	6.3	1.18	54.66	130.4	19.77	1.744
Ural	0.24	23.13	8.08	16.51	16.86	9.89	0.814
Siberia	0.3	74.8	7.56	50.24	60.24	10.93	0.93
Far-East	0.08	80.1	12.35	17.21	3.14	4.65	0.35

From table 11 it is seen, that all federal regions of the country have high potential for biomass resources. This level of biomass resources increases the opportunity for RE

[51] Derived by the author, based on information from Scheme and program of IES of Russia development for the period 2011-2017

[52] Derived by the author, based on information from (Bezroukikh, 2007, pp. 73, 78, 85, 196, 197).

technologies to be installed; first of all, because of the stable process of electricity production from biomass plants, it provides the necessary amount of electricity and power in to the market, whereas all nuclear PPs and most of the heat PPs provide demanded power in to the system and secure the load. (SO, 2011) Second, because biomass plants can be constructed in a combined cycle regime, in order to provide electricity and heat; 52% of planned heat PPs are CHP. It is planned to provide not only electricity in each region, but also heat (Federation, 2012).

The next option to satisfy planned installation is by hydro PP. Most of the regions have significant hydro resources, which can be used to provide a stable supply to the system and potentially replace the number of planned commissions by nuclear PP and CPP.

The last option to replace conventional resources is by solar PP. Currently it has low economic potential in the sector, and high investment costs in comparison with other technologies. The replacement of conventional PPs was made according to capital costs to each RE technology and availability of resources in each sector. Table 12 shows range of capital costs per kW of each technology.

Table 12: Capital costs of Conventional and RE power plants[53].

Type of plant	Capital costs, €/kW
Nuclear PP	2012.7
Heat PP	1213.7
Biomass CHP	3487.5
Biomass	2610
Biogas Agric. CHP	3642.5
Biogas Agric.	3420
Biogas Landfill CHP	1800
Biogas Landfill	1650
Biogas Sewage CHP	2975
Biogas Sewage	2850
Hydro medium scale	3600
Hydro small scale	1287.5
Wind onshore	1325
Solar	4312.5

[53] Derived by the author based on information from (Fraunhofer, 2011, S. 13) and (SO, 2011, pp. 82, 83).The capital costs of RE technologies were considered on the average level of listed prices. The capital costs of conventional technologies considered by aggregative measures of costs per unit.

The possible replacement in the North-West, Central, Volga and South regions is shown in the following results, represented in figure 30.

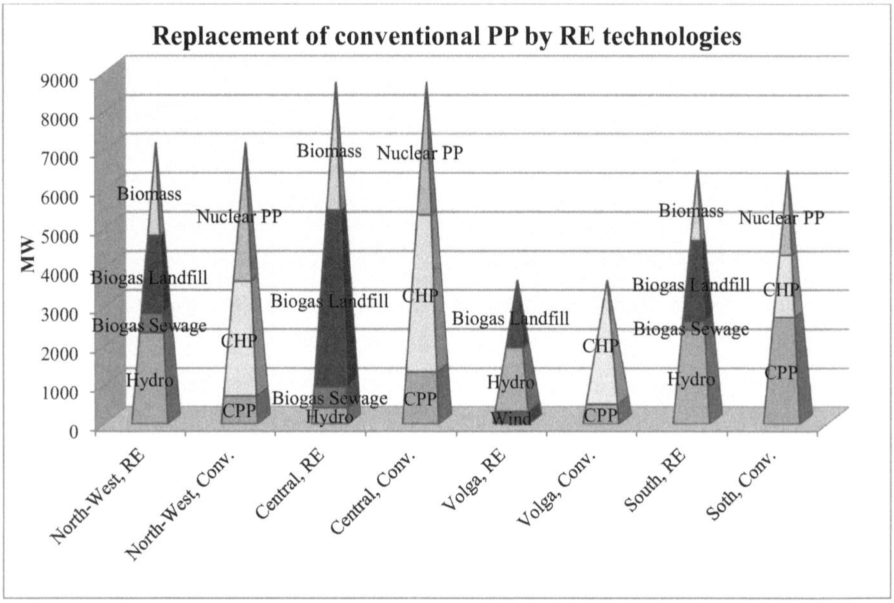

Figure 30: Replacement of conventional PP by RE technologies in the North-West, Central, Volga and South federal regions[54]

Figure 30 shows that North-West, Central and South regions plan to construct a high number of nuclear PPs, which amount is almost half of the planned construction in this sector. The Volga region is focused on CHP power plants with a small number of CPP. In all 4 regions there are enough biomass, biogas and hydro resources to replace conventional technologies. The North-West region will use half of the economic potential to install 7 GW of power plants by 2017; the other three regions will still have left 60-80% of the current economic potential after the installation of the required capacity of power plants.

A comparison of capital costs between conventional and RE technologies shows that the capital costs for RE technologies are considerably higher, by 21-34% in the North-West and South region and by 54% in the Central region.

---

[54] Derived by the author, based on information from (SO, 2011), (Fraunhofer, 2011), (Bezroukikh, Handbook of Renewable Energy Resources of Russia (Справочник по ресурсам возобновляемых источников энергии России и местным видам топлива), 2007)

The result of replacement in the Ural, Siberia and Far-East federal regions is represented in figure 31.

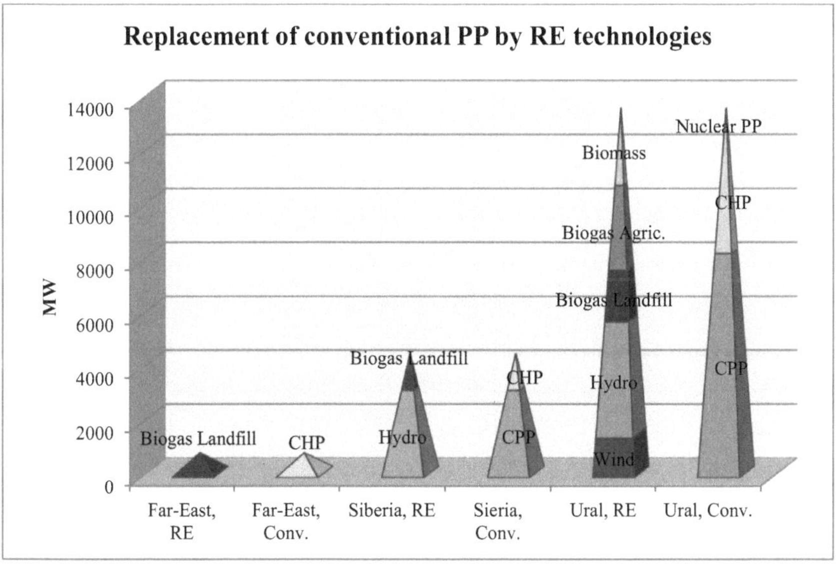

Figure 31: Replacement of conventional PP by RE technologies in the Ural, Siberia and Far East federal regions[55]

Figure 31 shows a high level of construction of CPP in the Ural and Siberia regions, with 880 MW of nuclear PPs in the Ural region. The lowest level of construction is planned in the Far East region. The greatest construction of power plants in the whole Russian IES is planned in the Ural federal region. This is explained by high growth of demand in that sector, with the construction of big power-consuming industries. At the same time, the Ural federal region has the lowest economic potential for RE technologies in the sector. To replace the necessary number of planned conventional technologies by 2017, it will consume all biomass, biogas, hydro and wind resources. The Siberia region will consume only 10.2% of its RE resources, and the Far East region even less – 2.3%.

[55] Derived by the author, based on information from (SO, 2011), (Fraunhofer, 2011), (Bezroukikh, Handbook of Renewable Energy Resources of Russia (Справочник по ресурсам возобновляемых источников энергии России и местным видам топлива), 2007)

The capital costs for RE technologies in the Ural region are higher than conventional technologies by 81%; the costs for Siberia and the Far East region are higher by 18% and 48% respectively. Detailed tables of calculation comparison can be found in Annex 3.

The high costs of RE technologies should not be considered as a negative factor for further implementation of these technologies. First of all, the calculations were made in order to satisfy the requirements of conventional PPs to provide electricity and heat. As can be seen in table 12, Biomass and Biogas technologies costs are higher for typical construction by 6.5% - 33.6%, and the considered costs are costs for European countries. If Russia uses its own facilities to produce and construct RE technologies, it will significantly reduce the capital costs of each RE technology.

At the same time, to calculate costs per unit of conventional technologies SO used typical standard costs for each kind of technology: nuclear, CHP, CPP. In reality, costs for construction of CHP, especially with usage of new modern technologies can be higher; the construction of nuclear PP has much higher uncompetitive features compared with RE technologies. The capital costs for construction do not include costs for decommissioning the power plant after plant lifetime expiration. Different sources estimate such costs to be around 20-40% of the capital costs of the plant (Institute, 2012) (Alekssenko I. R., Keisevich L. V., 2011).

Therefore, RE technologies are competitive with conventional power plants in the Russian energy sector, especially when operation and maintenance costs are compared, and costs for environmental protection, which would obviously be considerably higher for energy producers from conventional sources.

Operation and maintenance costs for RE technologies vary from 30-40 €/kW/year for wind, PV and hydro PPs to 84-175 €/kW/year for biomass and biogas PPs (Fraunhofer, 2011, p. 13). For heat PPs operation and maintenance costs vary from 90-300 €/kW annually (Lako P., Simbolotti G., Tosato G., 2010, p. 5). In addition, the yearly increase in fossil fuel costs for heat PP should also be considered in the cost calculations.

For nuclear PPs the cost of energy generation is approx. 1.23 €/kWh with a fuel cost of 0.53 €/kWh (Institute, 2012). This level seems to be low for energy production, but here should be added costs related to waste management; in total, during the whole lifetime period of nuclear PP, producers should spend approximately 30% of the capital costs of the plant to utilize the wastes of the plant (Institute, 2012), (Alekssenko I. R., Keisevich L. V., 2011). That means, the real capital costs of nuclear power plants are 60-70% higher than the estimated costs.

Furthermore, currently there is no effective technology for nuclear waste utilization, and what kind of additional impact it will have on the environment is hard to predict. Nuclear PPs construction in Russia is under the control of the government, and the government can decide not to risk harming the environment in the country and save huge quantities of financial sources for additional waste utilization.

All estimated figures represented above show a high competitive potential for RE technology development. Furthermore, the economic potential of RE resources was considered according to the level of prices for fossil fuel in the market, electricity and heat, equipment, transport service, labor payments. As the level of fossil fuel prices has a tendency to rise, it means that the economic potential of RE sources will only increase in the future.

However only a small part of the Russian energy sector is promoting RE technology development. And this part includes only government enterprises, which are intending to develop more than 4.5% RE technologies, but this is still far from reaching the significant economic potential for RE implementation. Currently, the main entity promoting RE is Rushydro. Within its strategy Rushydro considers commissioning new RE PPs (see chapter *Government and private enterprises*).

In addition to the Government Energy Strategy, Rushydro proposes more places and projects for RE development. Thus, it is promoting up to 2020 7% of wind PP development, instead of 0.3% proposed by ES, in the Volga region, Volgograd, Krasnodar, Murmansk, Novosibirsk and Primore region; 4% instead of 1.3% of biomass PPs mostly in the South, Volga and North-West regions; 0.5% of geothermal PPs in the South and Far East regions, Kamchatka and Kurily Islands; 2% of hydro PPs in Karelia, Siberia, the Far East and South regions; and 2% of solar PPs in the South region and Far East (Rizhinaschvili, 2011, p. 7). Figure 32 represents operating RE projects, and projects proposed by Rushydro for implementation.

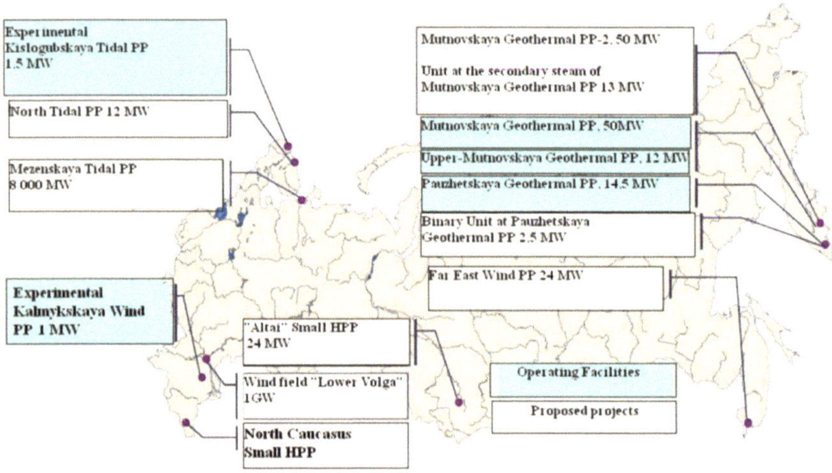

Figure 32: Proposed renewable energy projects for implementation[56].

Also Rushydro formulated a portfolio of prospective Small HPP projects in different regions of the Russian Federation (Rizhinaschvili, 2011, p. 24):

1. Central Region – 9 projects with installed capacities of 23 MW
2. South region – 230 projects with installed capacities 1288 MW
3. North-West region – 23 projects with installed capacity 134 MW
4. Volga region – 17 projects with installed capacity 135 MW
5. Ural region – 15 projects with installed capacities 41 MW
6. Siberia region – 56 projects with installed capacities 220 MW
7. Far East region – 60 projects with installed capacities 258 MW

Another entity, which is promoting RE development, is ESE. In its innovative program, ESE considers the development of all types of RE PP. In particular, it is considering construction of 3 wind PPs of different capacities and the development of wind energy technologies in the Yamalo-Nenezkii autonomous region from 2011 to 2020 (see table 13):

[56] Derived by the author, based on information from RizhinaschviliJ.I. presentation "Growth of renewable energy technologies usage is dominate tendency for energy sector development in the world" ("Рост использования возобновляемых источников энергии – доминирующая тенденция развития электроэнергетики в мире"), St. Petersburg, 2011, www.rushydro.ru

Table 13: Potential renewable energy projects in Far East[57].

Name	Years
Wind PP Labytangi town, 275 kW	2011-2018
Wind PP Bering Island, 550 kW	2011-2018
Wind PP Ust-Kamchatsk, 275 kW	2011-2018
Wind PP development at YamaloNenezkiiAutonomus region	2013-2020

Potential in the energy efficiency sector

The energy efficiency sector in Russia is broadly developing in all three spheres: generation, transmission and consumption.

Generation

As ES sets a target of 59% of installed capacities from heat PPs in 2030, generation companies owning heat PPs and engineering industries are focused on high efficiency technologies in electricity production from fossil fuel resources. Such technologies include the following types of new generation:

- Steam-turbine and gas-turbine plants with plants of high- and super high critical steam parameters
- Integrated gasification combined cycle plants
- Pressurized fluidized bed combustion plants

Currently, these technologies have already achieved 45% efficiency and low emissions of GHG. The steam-turbine plants have already achieved indexes of 545-580 °C and 24.5-30 MPa. These technologies are already in serial production in the engineering industry (Dub A. V., 2007).

By 2020, engineers and scientists plan to achieve indexes of 600 °C and 30MPa and efficiency of 60%. In 2030 this level should be increased to 740 °C and 35 MPa. It is also predicted that future technologies can achieve 70-80% efficiency (Industry, 2011).

The implementation of high efficiency technologies is broadly supported by the government. The President's decree from 07.07.2011 № 988 „Statement of priorities of science,

---

[57] Derived by the author, based on information from (ESE, 2009, p. 17)

technologies and techniques development in the Russian Federation and list of critical technologies of the Russian Federation" expresses high priority for such types of technologies. In addition, the Ministry of Industry and Trade approved the "Strategy of engineering industry development in the Russian Federation for the period 2010-2020 and for further development until 2030".

In its strategy the ministry has set the following tasks to be achieved in three periods:

1. By 2016:
    - provide domestic supply of auxiliary materials
    - commercial production of gas-turbines, steam-gas turbines of 60% efficiency
    - commercial production of steam-gas turbines with the production of 1200-1500 kWh/GCal and efficiency 85-86%
    - commercial production of steam-gas turbines with innercycle coal gasification and 50% efficiency
    - Commercial construction of Nuclear PP VVER-1200
2. By 2020:
    - Commercial production steam-gas turbines of 600–620°C and 30 MPa. And first construction of plants with parameters of 700-740 °C and 35 MPa
    - First demonstration construction of gas-turbine plant with efficiency of 70%
3. By 2030:
    - Commercial production of gas-turbine plant with efficiency of 70-75%
    - Commercial installation of carbon dioxide capture and storage equipment

The IEA has identified that 45% efficient technologies reduce emission of CO_2 from the basic level by up to 25%. And technologies with carbon dioxide capture and storage equipment increase the level of CO_2 emission reduction up to 90%. Furthermore, in case of further implementation of Russia's Kyoto protocol liabilities in the form of standard implementation or other forms, generation companies producing electricity from heat PPs can get additional support from Joint Implementation agreements, as the UNFCCC approved a consolidated methodology in September 2007 for new, grid-connected fossil fuel-fired power plants using less GHG-intensive technology that enables them to sell carbon offsets under the Kyoto Protocol's Clean Development Mechanism and Joint Implementation (IEA I. E., 2008, S. 24, 28)

Also the potential for energy efficiency generation is broadly supported by different research organizations in Russia, most of which are concentrating mainly on efficiency technologies in heat and nuclear PPs. For example, the research organization of technical physics developed an experimental plant of 350W, which uses hard oxide fuel elements to produce electricity. Temperature of electro-chemical reaction is at a level of 600-1000 °C; natural gas, petrol, kerosene, and diesel can be used for energy production. Efficiency of the plant reaches 90%. The innovation center "Skolkovo" has announced a new Russian-German enterprise, which plans to produce hard oxide fuel element plants from 1.5 kW till 20 MW. (Vesti, Vesti. Energetika, 2012).

The Siberian Government Industrialized University has developed a new mini PP, which is used for electricity and heat production and in the process produces waste from coal slack. Coal cleaning creates 7-10% of coal slack from the whole volume of coal concentrate. Scientists developed a special utility to dilute coal slack and create slurry coal fuel. This fuel burns in a swirling-type furnace and produces electricity and heat. Usage of a swirling-type furnace significantly decreases emissions of GHG (Vesti, 2012). PP is fully automated with minimum employee participation. An experimental PP was constructed with a capacity of 250 KW. Net costs of the plant will decrease with the construction of big capacities PPs. (Vesti, 2012)

Another example of new scientific technologies is the floating nuclear PP (FNPP). This type of power plant has already been put into serial production. The plant is composed of a non-self-propelled boat with two reactors KLT-40C. The installed capacity of the plant is 70 MW, heat capacity 140 GCal/h, lifetime period at least 36 years; scheduled repair should be provided every 12 years (Rosenergoatom, 2012). The first FNPPs are planned to be constructed in 2012 in Pevek city, Chukotka and Viluchinsk city, Kamchatka (Stroy-firms, 2011). These places are characterized by a lack of electricity and heat energy. The principal feature of FNPP is an installation only on the water surface; that means plants are restricted by coastal regions.

All these measures represent potential for energy efficiency in the engineering industry.

The following information will show energy efficiency potential from generation facilities. Specialists from the high school of economics, the institute of transition economy with the cooperation of ETSAP IEA and WWF of Russia developed a model of GHG emissions by big energy facilities in Russia. The model results show that PPs consuming fossil fuels emit 75% of the total CO_2 emissions and 50% from Russia's total GHG emissions. (Kokorin A. O., Safonov G. V., 2009).

Coal consumption produces the highest emissions of CO_2 – 87.5%; although the share of coal in the structure of fuel consumption of heat PP in Russia is 22.1% and gas is 71%, the remaining consumption is of black oil and other fuels. The reason for such high emissions by coal consumption is growth of the ash content in black coal and lignite. (Linev B. I., Rubinschtein Y. B., 2009).

The forecast for primary energy consumption in electricity production shows that in 2030 the share of coal will increase from 14.6% to 25.6% (Linev B. I., Rubinschtein Y. B., 2009). This figure should alert government and generation companies to make the increase of energy efficiency technologies a high priority.

To analyze the up to date situation in the generation part of the market represented by heat PPs and compare current status with proposed measures by SO, the information on heat PPs conditions was collected and consolidated from private enterprises. The focus on private enterprises was made because of the density of heat PPs in this area of the market. The information was collected from five wholesale generation companies, 14 territorial generation companies and the Far East Generation Company. (The list of existing heat PPs with information of date of construction and technical parameters can be found in Annex2.)

From consolidated information it was found that companies own 209 CHP and CPP power plants. The lifetime period of all types of heat PPs was assumed to be 40 years. By the analysis it was found that from 209 power plants 127 have been functioning for more than 40 years and have to be reinstalled soon. The expiry date of 14 power plants from 209 is in 2016; 24 power plants have to be reinstalled in the period between 2020 and 2030; and 44 are new generations in the sector and have a lifetime period till 2055.

The represented figures show a high potential for new energy efficiency technologies in the sector. As it was discussed above, current technologies of engineering industry provide 45% of efficient steam-turbine and gas-turbine plants, which reduce GHG emission by more than 25%. That means, 127 power plants currently have to be reinstalled which will increase the level of efficiency on average by 15% from the basic level and will significantly reduce the emission level of GHG.

The 14 PPs can be reinstalled by the PPs with 60% efficiency in 2016, according to the level of engineering industry development. The 24 PPs will be close to very high energy efficiency technologies in the period 2020-2030; and the modern 44 PPs can be completely re-equipped by innovative technologies of high efficiency with technologies of carbon dioxide capture and storage which will potentially reach a level of 90% emission reduction.

However, reinstallation of 127 PPs at once is not possible. It was calculated that 95 PPs from 127 have an expiry date before 2000 and have to be modernized or re-or uninstalled. The total capacity of 95 PPs is 17.36 GW. According to the SO scheme, generation companies plan for the period 2011-2017 modernization of 4.5 GW of heat PPs, reinstallation of nearly 3 GW and uninstallation of nearly 7 GW with a further exchange of 2.3 GW, that aggregates 16.8 GW.

That means that by 2017 generation companies will not cover only 0.56 GW of the necessary 17.36 GW. Also, these measures show that obsolete PPs will be renovated by 2017 and a major number of heat PPs in the sector will reach the level of modern power plants.

A more detailed calculation of new modern generation facilities' impact on the environment couldn't be made within the scope of this research. Further research should be conducted in order to estimate future impacts of innovative generation facilities on the environment.

Transmission

Energy efficiency measures in the transmission sector are encouraged by the development of the generation and consumption sector. The system operator in its "Scheme and program of IES of Russia's development from the period 2011-2017" (Scheme) establishes phased development of the transmission sector. There are some areas in the sector, where the transmission system has bottlenecks for sustainable transmission and distribution of electricity and power. The scheme established a priority list for the construction of transmission lines and sub-stations to reduce the amount of bottlenecks in the sector. Most such transmission lines and substations have standard capacity, and for some new heavy generation it is planned to construct high voltage transmission lines, which are under the modern conception of a super grid.

It is also planned to modernize and implement new innovative technologies in the system by collecting and transforming information, automatic regulation of frequency and intercharge of active power and systems of transient regime controlling in the energy system. All these technologies are considered the first step of a super grid implementation.

FGC and ESE have already considered all measures mentioned above in their investment and innovative programs. The IDGC by SO is considered a backward company in terms of fast and innovative development. Additionally, in July 2012 FGC and IDGC merged into a united company. The main reason for this decision was to initiate a faster development and operation of IDGC. Below is a description of the investment and innovative programs of ESE and FGC.

ESE has developed three main lines of innovative technologies, which are supposed to be implemented within the investment program:

1. Technologies of energy efficiency transmission, distribution and consumption. This includes (ESE, 2009, pp. 68-77):

    - Implementation of high voltage transmission lines of increased capacity, having low consumption of ferrous and non-ferrous metals and small right of way. All these features have to reach high ecological compatibility.
    - Insertion of HVDC lines, to transmit for a long distance and arrange nonsynchronous connection of energy systems
    - Implementation of static reactive power sources, such as static thyristor compensators, controlled shunt reactors in combination with capacitor banks, etc.
    - Implementation of heat-resistant wires with a high allowable frequency of current.
    - Implementation of wire for overhead power lines, for low voltage transmission lines up to 20 kV
    - Implementation of Aero-Z wire
    - Modernization of transmission protection devices
    - Smart Metering implementation
    - Etc.

2. Technologies of energy production with low environmental impact and energy efficiency (ESE, 2009, p. 72.73):

    - Exchanging the main parts of generation facilities into gas, with construction of gas-turbine and combined cycle gas-turbine plants of new generation (heat recovery of flue gases, flue gas recycling)
    - Thermal clamping units of new generation, with a steam temperature of 600 °C and performance ratio up to 90%
    - Construction of Geothermal PPs

3. Technologies of new renewable energy facilities. (See Chapter *Potential in the sector of renewable energies*)

Table 14 represents some of the investment projects, in which the described innovative technologies are supposed to be used.

Table 14: Innovative investment projects of "Energy systems of East" JSC[58].

Name	Years
Construction of SahalinCondencing PP-2	2013-2018
Reconstruction of South-Sahalin CHP-1	2013-2019
ReconstructionVladivostok CHP-2	2013-2019
Construction of energy bridge Khabarovsk region-Sahalin (generation, transmission)	2012-2016
Sahalin region wire construction and reconstruction	2013-2018
Construction and reconstruction of distribution lines of mainland part of Valdivostok city	2012-2018
Construction sub-station "Omchak"	2013-2018

FGC is considering development of new technologies within its innovative program for the period up to 2016, with prolongation up to 2020. In this program, among other things, the realizations of complex pilot projects are considered, with further implementation of successful projects in an overall investment program. For implementation of such projects the program aims to attract not only domestic funds, but also foreign investment.

Thus, the main measures of innovative development include:

Development of an **Intelligent power system with an adaptive active-network**. By its characteristics this network is similar to the concept of a Smart Grid. There are some technologies considered, which form the basis for a Russian Smart Grid development (Center, 2012):

- Super conductivity technologies
- Technologies of electricity accumulation
- Devices of current limitation
- Technologies of digital substations
- Technologies of alternative current management
- Electricity transmission by HVDC
- Technologies of monitoring and diagnosing electrical networks

Development of all these technologies is included in the Innovative program of FGC.

It is planned to implement 7 pilot projects of Smart Grid (Center, 2012):

1. Energy cluster “Elgaugol” (Elgin coal mine, Saha Republic, Yakutia)

[58] Derived by the author, based on information from (ESE, 2009)

2. Energy cluster “Vanino” (Khabarovsk region)
3. Capacity expansion of Primorsky krai transmission lines (Primorsky krai)
4. Energy cluster “Kola” (Kola peninsula)
5. Energy cluster “Komi” (Archangelsk region, Komi Republic)
6. Energy cluster “Big St. Petersburg circle“
7. Energy cluster “Small St. Petersburg circle”

Consumption

Russia’s energy efficiency consumption sector has high priority for broad development. The Government program “Energy savings and improvement of energy efficiency for the period till 2020”, described in the chapter *Potential measures in electricity consumption*, is under high implementation by different players of the market. First of all, by the government technological platform “Smart Russian energy system”, which mostly controls the process of program implementation in the budget sphere of the market and realization of federal law № 261 “About energy savings and increasing energy efficiency…” from 18. 11. 2009. The TP has been performing quarterly inspections of law and the program implementation in every federal and municipal region of the country (Agency, Russian energy agency, 2012).

Another player of the market, which is highly focused on energy efficiency programs implementation, is the innovation center “Skolkovo”. By 2011“Skolkovo” had established 84 companies focused on energy efficiency technologies research and development; from 84 companies 28 companies are working on energy efficiency in the industry sector, 7 companies in energy efficiency in buildings, 7 – energy efficiency materials, 12 – generation and efficient storage, the rest 30 companies are focused on efficient transmission, green energy production and other energy efficient technologies. In 2011 23 companies out of 84 got 26.6 million euro as a grant from the Skolkovo grant committee for further development of energy efficiency technologies (Skolkovo, 2012).

Currently, the achieved results cannot show a clear picture of program implementation, it is appropriate to investigate results after completion of the first phase of the program between late 2015 and early 2016.

4 MARKET CONSTRAINTS PREVENTING RENEWABLE ENERGY AND ENERGY EFFICIENCY DEVELOPMENT

During the research and analysis three main constraints preventing renewable energy and energy efficiency development were found: conventional resources, governmental strategies and programs and legislative framework and tariff system. All three types of constraint are going to be considered in the following chapter.

4.1 Conventional resources

One of the main constraints to a broad renewable energy development in Russia is the abundance of conventional resources. As explained in previous chapters, Russia's energy sector relies mainly on gas, coal and oil for its electricity production. The second most used resources are nuclear and hydro energy. The last resource for energy production is RE.
The tendency to use fossil fuel resources will remain a priority in the planning for decades. This becomes even clearer when we consider the vast quantities of fossil fuel reserves in the country, as outlined in table 15.

Table 15: Fossil fuel reserves in Russia[59]

	Oil, bn ton	**Gas, Trillion m^3**	**Coal, bn. ton**
Reserve to production ratio[60]	10.2	44.38	157
	20.3	84.1	more than 500 years

Oil reserves are the least interesting resource for electricity production, as heat PPs use only mazut (black oil), as a tertiary fuel for energy production and as a starting fuel. The most interesting resource for consideration is gas, as gas is the primary fuel for more than half of the heat PPs. Russia has the second largest resource of natural gas in the world after the Middle East. Potential (expected and perspective) reserves of gas in Russia are estimated at 151.3 trillion cubic meters, which is 40% of the world's reserves, although the proven reserves are approximately 24% of the whole potential or 44.38 trillion cubic meters.

[59] Derived by the author from BP Statistical review of world energy, June 2010, bp.com/statistical review

[60] Reserves-to-production (R/P) ratio – If the reserves remaining at the end of any year are divided by the production in that year, the result is the length of time that those remaining reserves would last if production were to continue at that rate.

Currently, nearly half of prospective reserves need further exploration work. (Dolgikh, 2010)

Coal reserves are generally divided into two types: anthracite and bituminous and sub-bituminous and lignite. There are 49 bn tonnes of anthracite and bituminous coal in the area, and the largest part consists of sub-bituminous and lignite – 107 bn tonnes. (Beyond Petroleum, 2010)
As can be seen, Russia has significant resources of gas and coal. With the plans to use new innovative technologies, as it was described in the chapter *Potential in the energy efficiency sector*, and availability of gas resources for more than 80 years and coal resources for more than 500 years at current usage rate, and the lifetime of a typical heat PP of 40 years, it provides great incentive for a continuous conventional innovative development of the sector.

4.2 Governmental strategies and programs

The most important and strong constraint for broad renewable energy development is that of governmental strategies and programs. "Energy strategy up to 2030" provides the framework for energy development in Russia. As we have seen from the description of the RE potential and calculations of new generation constructions from RE, Russia has significant resources for higher RE technology development than 4.5% in 2030. Despite these possibilities however, the ES has set a fixed and hardly changeable quantity of RE technologies up to 2030. It is important to set the exact place of ES among other strategies in the sector. The ES is the highest and most important program here. From the ES originate all further strategies in the sector. This is a critical point for RE technology development in Russia for the next 20 years. Figure 33 represents the ES placement in the framework.

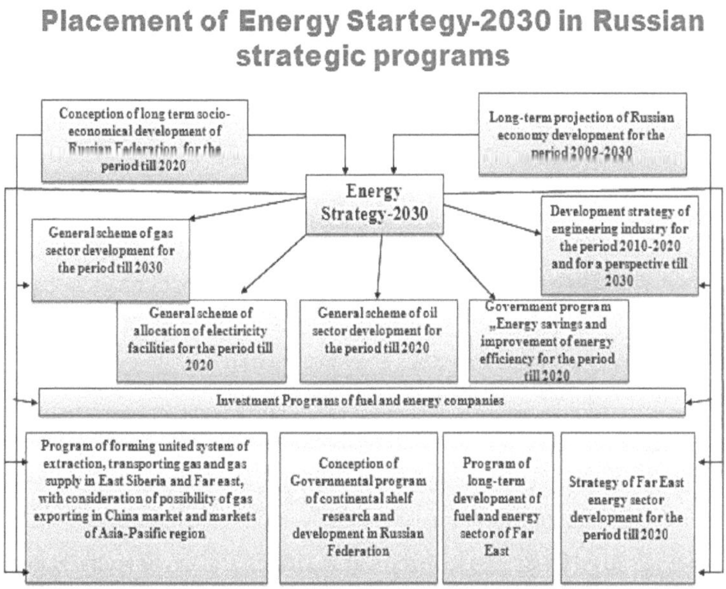

Figure 33: Placement of Energy Strategy for the period up to 2030 in Russian strategic programs[61]

At the same time the engineering industry of the country is currently incapable of installing the necessary 4.5% by 2030. As described in "Development strategy of the engineering industry for the period 2010-2020 and for a perspective to 2030", the industry needs broad support from the government to prepare the sector for RE technology penetration into the market. This support includes (Industry, 2011):

- Scientific activity aimed for RE technologies development
- Technical support, concerning creation of new types of generation equipment, accessories, etc.
- Organizational support, concerning integration of new generations in the united energy system of Russia
- Economic support, concerning economic incentives for production and penetration of RE technologies into the market.

This highlights that the engineering industry is completely unequipped for RE technology implementation. In terms of broad RE penetration it will highly increase capital costs of constructions, as all generation facilities and accessories would have to be imported from other countries. This makes investment in RE technologies unattractive.

[61] Derived by the author, based on information from (Government, 2009, pp. 1,2)

In addition, whole entities of the sector consider RE technologies only as auxiliary facilities to cover electricity demand. For example, the Ministry of Industry and Trade determines RE technologies as distributed generation less than 25 MW, which has to be installed close to consumers and cover a small amount of the sector's demand. While, main system generation is generation over 25 MW, located far from consumers, and proposed to cover the main part of the sector's demand. (Industry, 2011)
It is also still a widely held perception that RE technologies cannot be considered as reliable sources to cover demanded load and the system cannot be sustainably constructed based on using RE technologies as a general system generation. (Industry, 2011)

One of the other most important constraints, preventing green technologies developing, is the absence of RE technology promotion. During research and analysis it was found that in most of the strategies and programs generation companies consider RE technologies as one of the environmental facilities for further development, but this consideration doesn't include quantitative and qualitative types. Conventional resources and technologies on the one hand are broadly described by the type of technologies that will be expected to be developed; quantity and quality of these technologies etc. On the other hand, only small parts of the sector's entities pay attention to more detailed descriptions of RE technology development. One such entity is Rushydro. It promotes a detailed and broad development of RE technologies, in terms of different types of technologies. But this promotion stays within Rushydro and within TP "Renewable energy advanced technologies", established by Rushydro, FGC and others companies. It is not being popularized extensively in the country, and as a result, there is an absence of clear awareness by the electorate of the possibilities for the country's RE technologies. During the research, it was found that comprehensive promotion of RE technologies was largely missing, whereas the promotion of conventional resources was very much in the public media. Broad informative publications can be found, not only in the spheres related to the energy sector, but also in mass media, with emphasis on nuclear and gas energy development.

Russia's position regarding the Kyoto Protocol impedes further energy efficiency projects' implementation, especially in the construction of new innovative heat PPs in the sector with participation in Joint Implementation agreements. As a special representative of Russia, Bedritskiy A. on the 17th Conference of the Parties to the UNFCCC/7th Meeting of the

Parties to the Kyoto Protocol declared that Russia will fulfill its commitments in the first period by 2013, but will not commit to quantitative obligations in the second period of the Kyoto Protocol; although will proceed after commitment implementation to reduce anthropogenic emissions by 15-25%, but without mentioning an exact date for reduction (Bedritskiy, 2011). That shows that the country slows down active participation in the Kyoto Protocol and seeks an absolutely new type of cooperation. Probably, further successful support, such as Joint Implementation agreements of the Kyoto Protocol, that the sector will receive in terms of new universal climate agreement for the post-2012 period. Such universal agreement was also proposed by Russia on the 17th Conference of the Parties to the UNFCCC (Bedritskiy, 2011).

4.3 Legislative framework and tariff system

As outlined in the chapter *Current and future electricity supply*, the legislative framework for conventional generation is well developed after the restructuring program. For RE technologies it is only at the first stage for further development. It is seen, that the federal law № 35 „Electricity law" established a stable legal basis for further RE development. Afterwards, different executive acts were introduced to create a legal framework for RE technologies' entrance into the market. But these executive acts are unfinished legal frameworks, and these frameworks are not working in the sector.

As a first step to fulfill the RE legislative framework, the Government has to affirm the 2010 list of the following rules and regulations:

1. Regulation of RE premium calculation, which has to be added to equilibrium price of the market.
2. Regulation of the amount of obligatory purchase of electricity, produced by RE generators.
3. Rules of the cost calculation of electricity produced by RE on the wholesale and retail markets and in non-price zones.
4. Rules of cooperation of RE generators with grid companies and ensured suppliers.
5. Criteria of subsidizing grid connection of RE generators with an installed capacity of less than 25 MW.

These rules and regulations are the basis for the decision making process of investment. They establish the necessary investment appeal for RE technologies, especially in the sector, where green energy technologies are almost absent. Without these rules and regulations, investors, in terms of decision to install RE technology are forced to negotiate with other market players on a standard basis. This option is quite uncompetitive and unattractive.

The technical process for connection into the grid is an important aspect of cooperation between RE generators and grid companies. The grid operator has to be prepared to connect into the grid intermittent generators, in terms of wind plant or PV panels. Also, during the research no regulation options could be found for off-grid generation systems. Russia has many regions, where such systems could be installed, for example, the Far East region, where half of the generation facilities are mobile diesel power plants (ESE, 2012).

As mentioned earlier, there is no legislative framework for broad energy efficiency technologies development. This is a lack of a regulatory system in GHG mitigation. According to the Russian environmental legislation, CO_2 emissions for electricity generation facilities are not standardized and it is not obligatory for companies to pay for higher emissions than the set standard. Consequently, companies do not provide any payments for such emissions (Mosenergo, 2012).

In addition, this lack of regulation creates a niche for unrestrained emissions of GHG by generation companies – absence of standardization in obligatory manner of supplied coal by the rate of calorific value, outlet of coal-volatile matter, index of coal grindability and temperature of ash melting. Also, a technical condition, currently in force, doesn't contain parameter of mass concentration of mineral admixture in coal; that allows suppliers in the sector to provide crude coal. Realization of such regulations is very important for the electricity generation sector, as the level of coal generation in electricity production in 2030 will increase from its current 14.6% to 25.6% in 2030. (Linev B. I., Rubinschtein Y. B., 2009)

5 RECOMMENDATIONS AND CONCLUSION

5.1 Recommendations

As shown in the chapter *Potential in the sector of renewable energies* Russia has enough RE resources to cover its electricity demand. However, high availability of conventional resources and the chosen pathway of the government prevent the required RE technologies development. In the current climate it is not possible to overcome the constraints of the sector. Nevertheless to develop the sector on the settled 4.5% in 2030 or even more seems highly possible.

The first measure to achieve this in the near future is affirmation by the government of the rules and regulations as discussed in chapter *4*. This will create the legal basis for further RE technology development.

The second measure is to create additional incentives in the market. As discussed earlier, RE technologies have high development potential in remote areas. This means that the Russian government has to transfer liabilities of additional incentives for development from the federal government to regional development. Russia is a big country with different climatic zones and types of technologies, which can be good for one region but not appropriate for another. By this division of responsibilities Russia will provide each region with the tools to develop those technologies, which are suitable and appropriate for the specific region.
The types of incentives that can be adopted are similar to those, which are used in the countries with high RE technology development, e.g. European Union. These incentives include (Fraunhofer, 2011):

- Feed-in-tariff – is a fixed and guaranteed price paid to the eligible producers of electricity from renewable sources, for the power they feed into the grid.
- Feed-in-tariff premium – a guaranteed premium is paid in addition to the income producers receive for the electricity from renewable sources that is being sold on the electricity market.
- Quota obligation – the government creates a demand through imposing an obligation on consumers or suppliers to source a certain percentage of their electricity from renewable sources.
- Investment grants
- Tax exemptions
- Fiscal incentives

The first two incentives concern the tariff system in the market. Some kind of tariff premium in the market has already been proposed by the government but has so far not been affirmed and is still under consideration. Once the tariff premium is in place, it will encourage producers to invest in RE technologies. The feed-in-tariff system by itself has the good possibilities for remote areas of the country, where the government still establishes a tariff for all participants of the market or so-called non-price zones of the electricity market. This system will allow small producers to enter the retail market. It is better for each administrative and federal region to establish its own feed-in-tariff system, by giving priorities to those technologies, which have more potential in this area.

A quota obligation system has already been proposed by the government for the consumers, but still has to be affirmed. In addition to quota obligation for consumers, the government can establish quota obligations for electricity producers. It will encourage producers to build into every investment program of each company a certain quantity of RE technologies. This obligation has to be made according to the regional location of each company and levels of RE sources in each region.

These incentives are very appropriate, for both, the investors and producers of electricity in the sector, as well as the engineering industry, which, as mentioned above, has a very low level of RE technology production. With the establishment of such incentives in remote regions of the country, the government will increase promotion of RE technologies, reduce constraints and increase RE technology development.

Regarding energy efficiency constraints and overcoming measures, the government has to pay close attention to the coal industry development. As the research showed, electricity production from coal is the most polluted production in the whole sector. It depends not only on the quality of the PP, but also on the raw product supplied into the plant. Although, the Minister of Energy of the Russian Federation confirmed in its last operational meeting with the head of the government that the coal industry will be forced to achieve the targets of a long term development program for the coal industry for the period to 2030 and that it will reach a level of coal preparation of 60% in all regions of Russia, and 80% of the total amount of mined coal will be under sorting, coal preparation, deep treatment and gasification (Energy, 2012). This sphere of legislative framework has to be broadly improved and stay under the strict control of the government.

The energy efficiency sector can receive additional strong support in terms of further implementation of the Joint Implementation agreements of the Kyoto Protocol or similar agreements; as it provides good additional investment sources for generation companies which plan to reconstruct or build new heat PPs. This support will let generation companies consider expansive, but highly efficient and environmentally friendly equipment. But further development of this sector is highly dependent on decisions by Russia and the other participants of the Kyoto Protocol.

5.2 Conclusion

Analysis of conventional resources shows that Russia will continue development by the chosen pathway. The way the ES program is currently structured, until 2030 RE technology development will have low priority and will remain at the bottom of the investment programs of each company.

What kind of environmental impact it will have in 2030 and later it is hard to predict. It would not be right to say that the pathway of conventional development will have significant negative influence on the environment. In terms of fulfilment of all stages of the ES and programs of the engineering industry, the conventional energy has a big chance to achieve high results in protection of the environment while producing electricity and heat.

In case of a decision of significant transfer of conventional technologies to RE technologies it can put the whole economy of the country under great pressure and will hardly have an impact on energy security, as most areas of the country are united by IES, which transmits electricity from the strategically important nuclear, heat and hydro PPs. At the same time the engineering industry and scientific institutes have a good, yearly accumulated basis for the development of new innovative conventional technologies, where RE technologies development remains almost absent.

This means that Russia in terms of RE technology development has to invest quite significant resources for meaningful progress. The government, engineering industry and generation enterprises are clearly focused on conventional technology development. This also raises the question, how will the government implement all liabilities in every sphere of electricity generation, in order to fulfil targets to provide electricity and heat with minimum risk for the environment?

Nevertheless, RE technologies have high potential to be developed in remote regions of the country. Remote regions are a good market niche for RE technologies development, because these parts of the country have enormous RE resources, and some of them have a lack of electricity supply, e.g. areas of the Far East and northern parts of Ural and Siberia; there are areas of low population density, where RE facilities can be installed instead of high cost transmission systems from conventional generations.

To develop this sector of the market the government has to fulfil the legislative framework for RE technologies by affirmation of planned regulations and has to establish additional financial incentives through regional government institutions. Such financial incentives should not strain the government budget in order to develop conventional systems and RE technologies at the same time, but will create technology appeal for private investment in the sector.

In the end, additional research is required, in terms of comparison of the two pathways for development: conventional technologies development with consideration of high efficient technologies implementation and RE technologies development; in order to calculate indexes of GHG emissions by the two pathways in the country, and to see the influence of the two pathways on the environment. At the same time, it will be equally important to consider how the implementation of the different pathways will influence the economic situation in the country, as well as energy security.
Additional research can be done in terms of the results of financial incentives program implementation, in order to see the effects of this program on the further development of the RE technologies penetration in remote areas.

6 BIBLIOGRAPHY

Agency. (2012, August 06). *Russian energy agency*. Retrieved August 06, 2012, from rosenergo.gov.ru: http://rosenergo.gov.ru/activity/analitics/energyefficiency/

Agency. (2012). *Russian Energy Agency*. Retrieved May 15, 2012, from rosenergo.gov.ru: http://rosenergo.gov.ru/upload/Brochure_2012_04.pdf

Alekssenko I. R., Keisevich L. V. (2011). *Educational portal "Youth year"*. Retrieved July 20, 2012, from www.godmol.ru: http://www.godmol.ru/ekologija/183-kilovat-chas.html

Austria. (2012). *The Development Bank of Austria*. Retrieved May 15, 2012, from http://www.oe-eb.at: http://www.oe-eb.at/en/projects/pages/default.aspx

Bedritskiy, A. (2011, December 08). *Statement of the Advisor to the President of the Russian Federation to the 17th Conference of the Parties to the UNFCCC/7th Meeting of the Parties to the Kyoto Protocol*. Durban, South Africa.

Beyond Petroleum, (2010, June). *BP Statistical review of world energy*. Retrieved from bp.com: bp.com/statisticalreview

Bezroukikh, P. (2007). *Handbook of Renewable Energy Resources of Russia (Справочник по ресурсам возобновляемых источников энергии России и местным видам топлива)*. Moscow, Russia.

Center. (2012). *R&D Center for Power Engineering; Advanced Russian engineering research and development Center*. Retrieved May 2012, from http://www.ntc-power.ru: http://www.ntc-power.ru/ies-aas/pilot_projects/

Center. (2012). *R&D Center for Power Engineering; Advanced Russian engineering research and development Center*. Retrieved May 2012, from www.ntc-power.ru: http://www.ntc-power.ru/ies-aas/

CIA. (2012). *Central Intelligence Agency*. Retrieved 03. 31. 2012, from Central Intelligence agency: https://www.cia.gov/library/publications/the-world-factbook/geos/rs.html

Council, M. (2011). *"Market Council" noncommercial partnership*. Retrieved April 20, 2012, from www.np-sr.ru,: http://www.np-sr.ru/norem/information/russianelectricity/

Dolgikh, A. (2010). *Dolgikh Natural gas research*. Retrieved June 15, 2012, from dolgikh.com: http://dolgikh.com/index/0-11

Dub A. V., Schaschnov S. A. (2007). *Инновационные приоритеты в энергетическом машиностроении: опыт отраслевого Форсайта*; Форсайт №3 (3) 2007.

DVGK. (2012). *"Far East Generation Company" JSC*. Retrieved July 15, 2012, from http://www.dvgk.ru:
http://www.dvgk.ru/?redirect=search&ed_search=%F2%EE%EF%EB%E8%E2%EE

Ecoclub. (2007). *Nature of South Siberia and its protectors*. Retrieved April 25, 2012, from ecoclub.nsu.ru: http://ecoclub.nsu.ru/altenergy/images/karta2.gif

ESE. (2012). *"Energy systems of East" JSC*. Retrieved May 12, 2012, from www.rao-esv.ru: http://www.rao-esv.ru/harakteristika-energosistemy-dfo

ESE. (2009). *"Программа иновационного развития ОАО «РАО ЭС Востока»"*; "Energy Systems of East" JSC. Russia.

Federation, (2012). *Ministry of Energy of Russian Federation*. Retrieved May 12, 2012, from minenergo.gov.ru:
http://minenergo.gov.ru/activity/powerindustry/powersector/structure/

Federation, (2012, August 06). *Глава Минэнерго России доложил на совещании у Председателя Правительства РФ о перспективах развития угольной промышленности*. Retrieved August 09, 2012, from Ministry of energy of Russian Federation: www.minenergo.gov.ru

FGC. (2012). *Federal grid company of Integrated Energy Systems*. Retrieved May 10, 2012, from www.fsk-ees.ru: http://www.fsk-ees.ru/about/structure/

FGC. (2012, July). Program for investors, "Federal Grid Company of Integrated Energy Systems" JSC, Russia.

Fraunhofer. (2011, January 2). *Financing Renewable Energy in the European Energy Market;* Fraunhofer, Ecofys, Ernst&Young, Technische Universität Wien.

Fund, T. F. (2004, July). *«Renforcement de capacites et appui a la mise en place d'un fonds dedie a l'efficacite energetique en europe de l'est »*. Retrieved May 15, 2012, from www.ffem.fr:
http://www.ffem.fr/webdav/site/ffem/shared/ELEMENTS_COMMUNS/U_ADMINISTRATEUR/3-PROJETS/fiches%20communication/CZZ1295-072005-Efficacit%C3%A9Energ%C3%A9tique-Europe%20de%20l'Est.pdf

Governemnt. (2009, October 17). *Decree № 823 "О схемах и программах перспективного развития электроэнергетики"*, Russian Federation Government. Russia.

Government. (2009, November 13). *Energy strategy of Russia for the period up to 2030,* Russian Federation Government. Russia.

Government, (2010, December 27). *Energy savings and energy efficiency improvement for the period till 2020*. Russian Federation Government. Russia.

Gromov, A. (2008). *Russian energy strategy for the period up to 2030: the Way to the Innovative Energy Development.* Workshop «Energy technology collaboration: global challenges and coordinated global response», (p. 7). Moscow.

IDGC, (2012). *"Holding of inter-regional distribution grid companies" JSC.*
Retrieved May 15, 2012, from www.holding-mrsk.ru: http://www.holding-mrsk.ru/clients/territory/, http://www.holding-mrsk.ru/sites/, http://www.holding-mrsk.ru/about/facts/spravka/

IEA. (2011). *International Energy Agency.* Retrieved April 20, 2012, from http://www.iea.org/stats/pdf_graphs/RUTPESPI.pdf

IEA. (2011). *International Energy Agency.* Retrieved April 20, 2012, from www.iea.org: http://www.iea.org/stats/pdf_graphs/RUTPES.pdf

IEA. (2011). *International Energy Agency.* Retrieved April 20, 2012, from www.iea.org: http://www.iea.org/stats/pdf_graphs/RUELEC.pdf

IEA. (2012). *International energy agency.* Retrieved April 20, 2012, from www.iea.org: http://www.iea.org/stats/pdf_graphs/RUELEC.pdf

IEA, I. E. (2008, February). *"Clean coal technologies, Accelerating Commercial and Policy Drivers for Deployment"*, International Energy Agency.

Industry, (2011). *Development strategy of engineering industry for the period 2010-2020 and for a perspective till 2030*; Ministry of Industry and Trade of RF. Moscow, Russia.

Institute, N. E. (2012). *Nuclear Energy Institute.* Retrieved July 20, 2012, from www.nei.org: http://www.nei.org/resourcesandstats/nuclear_statistics/costs/

Karabanov, S. (2001). *The Prospects for Photovoltaic Development in Russia*, World Renewable Energy Network.

Kokorin A. O., Safonov G. V. (2009, April). *Можно ли просчитать развитие энергетики России на 15-20 лет вперед, чтобы составить надежный сценарий изменения выбросов парниковых газов?* "Мировая энергетика" №4 (63) .

Lako P., Simbolotti G., Tosato G. (2010, May). *Combined heat and power*, Energy Technology System Analysis Program (ETSAP), IEA, Technology brief E04-May 2010, www.etsap.org.

Linev B. I., Rubinschtein Y. B. (2009). *Роль стандартизированного угольного топлива в реализации программы чистой угольной энергетики*. Russia: ФГУП "Институт обогазения твердого топлива".

Martynov, A., Artukhov, V., Vinogradov, V. (1997). *Гидроэнергетические ресурсы и размещение ГЭС*; www.sci.aha.ru/RUS/wadb6.htm. Russia.

Mosenergo. (2012). *"Mosenergo" JSC*. Retrieved July 20, 2012, from www.mosenergo.ru: http://www.mosenergo.ru/catalog/2041.aspx

Nilsson, S. and Shvidenko, A. (1998). *Is sustainable Development of the Russian Forest Sector Possible*; UIFRO Occassional Paper No.11; http://www.iufro.org/publications/series/occasional-papers/.

OGC-1. (2011). *First generation company of wholesale electricity market*. Retrieved July 10, 2012, from www.ogk1.com: http://www.ogk1.com/company/geography/

OGC-2. (2010). *Second generation company of wholesale electricity market*. Retrieved July 10, 2012, from www.ogk2.ru: http://www.ogk2.ru/rus/about/info/

Povarov O.A., Britvin O.V., Nikolski A.I., Tomarov G.V., Vasiliev G.P. (2002). *Use of the Earth heat for local heat supply (Использование тепла земли для локального теплоснабжения).*

ProFinance(2012). *ProFinance Service.* Retrieved July 25, 2012, from www.forexpf.ru/currency_eur.asp/

Rizhinaschvili, J. (2011). *Рост использования возобновляемых источников энергии – доминирующая тенденция развития электроэнергетики в мире*. St. Petersburg.

Rosenergoatom. (2012). *Rosenergoatom*. Retrieved July 25, 2012, from www.rosenergoatom.ru: http://www.rosenergoatom.ru/wps/wcm/connect/rosenergoatom/site/about/development/floating_npp/

Rushydro. (2012, March 31). *Rushydro*. Retrieved May 15, 2012, from www.rushydro.ru: http://www.rushydro.ru/investors/capital/structure

Rushydro. (2012). *Rushydro.* Retrieved May 10, 2012, from www.rushydro.ru: http://www.rushydro.ru/company/strategy,
http://www.rushydro.ru/company/invest/investprogramm/2011

Service. (2012, March 30). *Decree of FTS № 228-e*; The Federal TariffService. Russia.

Skolkovo, F. (2012, June 20). *Year report - 2011*; http://community.sk.ru/press/b/results/archive/2012/06/20/godovoy-otchyot-2011.aspx. Skolkovo, Russia.

SO. (2011, August 29). *Scheme and program of IES of Russia development for the period 2011-2017*; "System operator of Integrated energy system" JSC. Moscow, Russia.

SO. (2012). *System operator of Integrated energy system; "System operator of Integrated energy system" JSC.* Retrieved May 2, 2012, from so-ups.ru: http://so-ups.ru/index.php?id=about

Starkov A., Landberg L., Bezroukikh P.P.,Borisenko M.M. (2000). *Russian Wind Atlas;* Riso National Laboratory and The Russian-Danish Institute for Energy Efficiency. Russia, Denmark.

Stroy-firms. (2011, July 5). *Stroy-firms.ru*. Retrieved August 02, 2012, from www.stroy-firms.ru: http://www.stroy-firms.ru/articles/4654.htm

TGC-1. (2012). *First territorial generation company*. Retrieved July 10, 2012, from www.tgc1.ru: http://www.tgc1.ru/production/electricity-production/

TGC-10. (2012). *Fortum*. Retrieved July 15, 2012, from http://www.fortum.ru: http://www.fortum.ru/company/production/

TGC-11. (2012). *Territorial generation company №11*. Retrieved July 15, 2012, from http://www.tgk11.com: http://www.tgk11.com/production/harakteristiki/tomsk/

TGC-12. (2011). *"Kuzbassenergo" JSC*. Retrieved July 15, 2012, from http://www.kuzbassenergo.ru: http://www.kuzbassenergo.ru/contact/1/company.php

TGC-13. (2012). *Eniseiskaya TGC*. Retrieved July 16, 2012, from http://www.tgk13.ru: http://www.tgk13.ru/about/structure/abtec/

TGC-14. (2012). *Territorial generation company №14*. Retrieved July 15, 2012, from http://www.tgk-14.com: http://www.tgk-14.com/company/bur_gen/

TGC-2. (2011). *Territorial generation company №2*. Retrieved July 10, 2012, from tgc-2.ru: http://tgc-2.ru/about/general.shtml

TGC-3. (2012). *Mosenergo*. Retrieved July 11, 2012, from http://www.mosenergo.ru: http://www.mosenergo.ru/catalog/291.aspx

TGC-4. (2012). *Generation company - Cvadra*. Retrieved July 11, 2012, from http://www.tgk-4.ru: http://www.tgk-4.ru/structure.html

TGC-5. (2012). *Territorial generation company №5*. Retrieved July 11, 2012, from http://www.tgc5.ru: http://www.tgc5.ru/about_rus.html

TGC-6. (2012). *Territorial generation company №6*. Retrieved July 12, 2012, from http://www.tgc6.ru: http://www.tgc6.ru/index.php?id=5

TGC-7. (2012). *Volga territorial generation company*. Retrieved July 12, 2012, from http://www.votgk.ru: http://www.votgk.ru/company/power/

TGC-9. (2012). *Territorial generation company №9*. Retrieved July 12, 2012, from http://www.tgk-9.ru: http://www.tgk-9.ru/filials_rus.html

UES. (2012, March 16). *Inter RAO UES Generating Assets: Target Shareholding Structure;* "Inter RAO UES" JSC; http://www.interrao.ru/upload/docs/Reorg_pres_eng.pdf. Russia.

UES. (2005). *RAO UES of Russia*. Retrieved April 17, 2012, from www.rao-ees.ru: http://www.rao-ees.ru/en/

UES, I. R. (2012, February). *Inter RAO UES*. Retrieved May 12, 2012, from www.interrao.ru: http://www.interrao.ru/en/investors/acapital/capitstr/

UES, I. R. (2012, February). *Inter RAO UES*. Retrieved May 20, 2012, from www-interrao.ru: http://www.interrao.ru/activity/investing/

Vesti, E. (2012, June 10). *Vesti, Energetika*. Retrieved July 25, 2012, from www.vesti.ru: http://www.vesti.ru/videos?vid=425686&cid=1219

Vesti, E. (2012, July 07). *Vesti. Energetika*. Retrieved July 26, 2012, from www.vesti.ru: http://www.vesti.ru/videos?vid=431807&cid=1219

WGC-4. (2011). *E.ON AG*. Retrieved July 10, 2012, from www.eon-russia.ru: http://www.eon-russia.ru/RU/SitePages/CompanyStructure.aspx

WGC-5. (2012). *Enel*. Retrieved July 10, 2012, from www.ogk-5.com: http://www.ogk-5.com/power_plants/map/nevinnomyssk_gres/pgu_410/

7 APPENDIX

Appendix 1: Renewable energy sources potential of the Russian Federation.

Table 16: Solar energy potential in Russia (Bezroukikh, 2007, p. 73)

Name of Federal region	Gross potential, PWh	Technical potential, TWh			Economic potential, GWh		
		Heat	Power	Total	Heat	Power	Total
	Solar potential						
Central	987.39	4700.9	402.4	5103.3	5594	351.23	5945.2
North-West	2072.47	7729.3	930.4	8659.7	144.21	179.1	323.31
South	1171.14	6608.2	487.3	7095.5	7908.4	136.07	8044.5
Volga	1637.5	7779.3	678.03	8457.3	6140.6	431.47	6572.1
Ural	2507.43	7674.6	1043.2	8717.9	1698	243.07	1941.1
Siberia	7815.36	33741	3252.9	36994	4675.3	302.38	4977.6
Far-East	9457.52	33564	3935.6	37500	1465.4	76.76	1542.1
Total by the country	**25648.81**	**101797**	**10730**	**112527**	**27626**	**1720.08**	**29346**

Table 17: Hydro energy potential in Russia (Bezroukikh, 2007, p. 78)

Name of Federal region	Gross potential, TWh	Technical potential, TWh	Economic potential, TWh
	Hydro potential		
Central	8.41	2.91	1.57
North-West	54.55	15.02	8.89
South	160.6	18.8	10.3
Volga	35	11.4	6.3
Ural	135	42.1	23.13
Siberia	435.06	135.8	74.8
Far-East	451.7	145.8	80.1
Total by the country	**1280.32**	**371.83**	**205.09**

Table 18: Wind energy potential in Russia (Bezroukikh, 2007, p. 85)

Name of Federal region	Gross potential, TWh	Technical potential, TWh	Economic potential, TWh
	Wind potential		
Central	28717	71.79	0.36
North-West	173034	432.58	2.16
South	70633	176.58	0.88
Volga	94502	236.26	1.18
Ural	646795	1617	8.08
Siberia	605192	1513	7.56
Far-East	987762	2469.4	12.35
Total by the country	**2606635**	**6516.61**	**32.57**

Table 19: Biomass energy potential in Russia (Bezroukikh, 2007, p. 197)

Name of Federal region	Gross potential, TWh	Technical potential, TWh	Economic potential, TWh
	Biomass potential		
Central	364.83	219.57	138.51
North-West	571.963	132.465	77.22
South	323.779	310.75	188.87
Volga	600.805	362.27	206.67
Ural	457.87	98.97	44.08
Siberia	1918.25	362.85	122.7
Far-East	1200.216	146.54	25.35
Total by the country	**5437.713**	**1633.415**	**803.4**

Appendix 2: Heat PPs of the private sector of the market.

Table 20 represents list of heat PPs owned by private generation companies with information of date of construction and technical parameters.

Table 20: Heat PPs of the private sector of the market

Name of the station	Installed electricity capacity, MW	Installed heat capacity Gcal/h	Year of construction, reconstruction	Expire date	Type of fuel by relevancy[62]
WGC-1 (OGC-1, 2011)					
Verchnetagilskaya CPP	1497	480	1950–1964	2004	gas, coal
Iriklinskaya CPP	2400	121	1970–1979	2019	gas
Cashirskaya CPP	1910	399	1922–1968, 2009	2049	gas, coal
Permskaya CPP	2400	420	1986, 1987, 1991	2031	gas
Nizhnevartovskaya CPP	1600	758	1993, 2003	2043	gas
Urengoiskaya CPP	24	410	1980	2020	gas
WGC-2 (OGC-2, 2010)					
Surgutskaya CPP-1	3280	n/a	1972-1983	2023	gas
Ryazanskaya CPP	3070	n/a	1975-1981	2021	coal, gas
Ctavropolskaya CPP	2400	220	1975	2015	gas, black oil
Novocherkasskaya CPP	2112	75	1965-2009	2049	coal, gas, black oil
Kkirishskaya CPP	2100	826	1969-1979	2019	gas, black oil
Troizkaya CPP	2059		1960-1976	2016	coal, black oil
Krasnoyarskaya CPP-2	1250	976	1961-1983	2023	coal, black oil
Cherepovezkaya CPP	630	39	1976-1978	2013	coal, gas
Serovskaya CPP	538	-	1954-1959	1999	n/a
Pskovskaya CPP	430	-	under construction		gas
Adlerskaya Heat PP	360	-	2010	2050	n/a
E.ON Russia (WGC-4) (WGC-4, 2011)					
Berezovskaya CPP	1600	n/a	1987-2013	2053	coal
Smolenskaya CPP	630	n/a	1978, 1985	2025	coal, gas
Curgutskaya CPP-2	5597	n/a	1985-1986, 2011	2051	gas
Shaturskaya CPP	1493	n/a	1925, 1978, 2010	2050	gas
Yaivinskaya CPP	1025	n/a	1925, 1978, 2011	2051	gas, coal
Enel (WGC-5) (WGC-5, 2012)					
Reftinskaya CPP	3800	n/a	1970	2010	coal
Sredneuralskaya CPP	1182	n/a	1936	1976	gas, black oil
Konakovskaya CPP	2500	n/a	1965,1969	2009	gas, black oil
Nevinnomysskaya CPP	1670	-	1960-1970, 1972, 1997-2007	2047	gas, black oil
TGC-1 (TGC-1, 2012)					
Apatitskaya CHP	323	735	1959	1999	coal

[62] The first type of fuel given constitutes 85-90% of the fuel usage; the following types are used as additional fuel or fuel to heat up power plants.

Name of the station	Installed electricity capacity, MW	Installed heat capacity Gcal/h	Year of construction, reconstruction	Expire date	Type of fuel by relevancy[62]
Murmanskaya CHP	12	1111	1934	1974	black oil
Petrozavodskaya CHP	280	689	1976	2016	gas, black oil
Avtovskaya CHP	321	1849	1956	1996	gas, black oil
Vasileostrovskaya CHP	135	1213	1932	1972	gas, black oil
Vyborgskaya CHP	278	1110	1954	1994	gas, black oil
Dunrovskaya CHP	142	185	1933	1973	gas, coal, black oil
Pervomaiskaya CHP	524	1715	1957	1997	gas, black oil
CHP - 5 (Krasnyi Oktyabr)			1922	1962	gas, black oil
Pravoberezhnaya CHP	180	1120	2006	2046	gas, black oil
Severnaya CHP	500	1188	1975	2015	gas, black oil
Zentral'nay CHP	75.5	1340	1999	2039	gas, black oil
Yuzhnaya CHP	1175	2480	1978	2018	gas, black oil
TGC-2 (TGC-2, 2011)					
Archangelskaya CHP	450	1358	1970-1979, 2011	2051	gas, black oil
Severodvinskaya CHP-1	188.5	679	1941	1981	coal
Severodvinskaya CHP-2	410	1105	1976, 2011	2051	gas, black oil
Vologodskaya CHP	34	582	1956	1996	gas
mini CHP "Belyi ruchei"	6	26	2006	2046	wood, peat
Costromskaya CHP-1	33	450	1930	1970	coal, blackoil, peat
Costromskaya CHP-2	170	811	1974	2014	coal, gas, black oil
Sharinskaya CHP	21	388	1965	2005	black oil, peat
Novgorodskaya CHP	200	542	1968, 2009	2049	gas, coal
Tverskaya CHP-1	17	178	1912, 1935, 1943	1983	gas, black oil
Tverskaya CHP-3	170	694	1973	2013	gas, coal, black oil, peat
Tverskaya CHP-4	88	620	1949, 1952	1990	gas, black oil, peat
Vyshnevolozkaya CHP	4	124	1950	1990	gas, black oil
Yaroslavskaya CHP-1	131	697	1934	1974	gas, black oil
Yaroslavskaya CHP-2	325	1185	1955, 1970, 1972	2012	gas, coal
Yaroslavskaya CHP-3	295	1477	1961	2001	gas, black oil
TGC-3 (Mosenergo) (TGC-3, 2012)					
Gov. PP (Smidovich)	86	951	1897-2001	2041	n/a
CPP-3 (Klasson)	589	408	1914	1954	gas, black oil
CHP-8	605	2192	1930	1970	gas
CHP-9	210	560	1933	1973	gas, black oil
CHP-11	330	1011	1935-1993	2033	gas, black oil
CHP-12	412	2043	1941-1981	2021	gas, black oil
CHP-16	360	1484	1963	2003	gas
CHP-17	192	712	1953	1993	coal, gas
CHP-20	730	2400	1952	1992	gas, black oil
CHP-21	1800	4958	1960-1995	2035	gas
CHP-22	1310	3606	1960-2008	2048	gas, coal
CHP-23	1420	4530	1966-2008	2048	gas, black oil
CHP-25	1370	4088	1976	2016	gas, black oil
CHP-26	1840	4214	1979-2011	2051	gas, black oil

Name of the station	Installed electricity capacity, MW	Installed heat capacity Gcal/h	Year of construction, reconstruction	Expire date	Type of fuel by relevancy[62]
CHP-27	1060	1876	1996-2008	2048	gas
TGC-4 (TGC-4, 2012)					
Belgorodskaya CHP	60	360.4	2005	2045	gas, blak oil, coal
Gubkinskaya CHP	46	201	1954	1994	gas, blak oil, coal
CHP "Luch"	60	62.4			gas, blak oil, coal
Bryanskaya CPP	38	70	1931	1971	gas, black oil
Klinzovskaya CHP	12	165.8	1937	1977	gas, black oil
Voronezhskaya CHP-1	168	1269	1933	1973	gas, blak oil, coal
Voronezhskaya CHP-2	127	785	1957	1997	gas, blak oil, coal
Kaluzhskaya CHP	12	70	1948	1988	gas
Kurskaya CHP-1	202	1097	1955	1995	gas, blak oil, coal
Kurskaya CHP-4	4.8	395	1934	1974	gas, blak oil, coal
Lipezkaya CHP-2	515	1002	1978	2018	gas, blak oil, coal
Elezkaya CHP	70	389.5	1955	1995	gas, blak oil, coal
Dankovskaya CHP	10	152	1961	2001	gas, blak oil, coal
Tambovskaya CHP	235	947	2008	2048	gas, blak oil, coal
Kotovskaya CHP	80	198	1960,1975,1985,1992	2032	gas, blak oil, coal
Orlovskaya CHP	330	855	1946	1986	gas, black oil
Livenskaya CHP	12	240	1958	1998	gas, black oil
Dyagilevskaya CHP	110	423	1959	1999	gas, coal, black oil
Smolenskaya CHP-2	275	774	1973-1975, 1982	2022	gas, blak oil, coal
Dorogobuzhskaya CHP	128	442.2	1957	1997	gas, blak oil, coal
Shekinskaya CPP	400	-	1950	1990	gas, coal
Novomoskovskaya CPP	246	698	1934	1974	gas, coal
Aleksinskaya CHP	102	215	1941	1981	gas, coal
Eferemovskaya CHP	160	720	1933,1935-1939	1979	gas, coal
TGC-5 (TGC-5, 2012)					
CHP-1	929.3	3369	1934	1974	gas, coal, peat, black oil
CHP-3			1942	1982	gas, coal, peat, black oil
CHP-4			1963	2003	gas, coal, peat, black oil
CHP-5			1981	2021	gas, coal, peat, black oil
TGC-6 (TGC-6, 2012)					
Vladimirskaya CHP-1	n/a	n/a	1928	1968	peat, gas
Vladimirskaya CHP-2	406.5	1161	1958-1995	2035	gas
Penzenskaya CHP	385	1168	1944	1984	gas, black oil
Kuznezkaya CHP-3	4	200	1933	1973	gas
Penzenskaya CHP -2	16	335.5	1957-1990	2030	gas, black oil
Saranskaya CHP-1		120	1933	1973	gas
Saranskaya CHP-2	340	778	1958-1999	2039	gas, black oil
Alekseevskaya CHP-3	9	73	1955	1995	gas
Ivanovskaya CHP-1	12	129.4	1928	1968	n/a
Ivanovskaya CHP-2	181	943.1	1954-1969	2009	n/a
Ivanovskaya CHP-3	330	1076	1977-1991	2031	n/a
Novogorkovskaya CHP	305	942	1956-1991	2031	n/a
Nizhegorodskaya CPP	144	566	1925-1984	2024	gas

Name of the station	Installed electricity capacity, MW	Installed heat capacity Gcal/h	Year of construction, reconstruction	Expire date	Type of fuel by relevancy[62]
Sormovskaya CHP	340	1006	1975	2015	gas, black oil
Igumnovskaya CHP	87	497	1939	1979	n/a
Dzerzhinskaya CHP	580	1474	1961	2001	gas, black oil
TGC-7 (TGC-7, 2012)					
Samarskaya CHP	440	2054	1969	2009	gas, black oil
Samarskaya CPP	61	1827	1900	1940	gas, black oil
Bezymyanaya CHP	169.7	1445	1941	1981	gas
Toljatinskaya CHP	620	2173	1960, 1968, 1972, 1975	2015	gas
VAZ CHP	1172	3903	1967, 1969	2009	gas, black oil
Novokuibyshavskaya CHP-1	144	624	1951-1956	1996	gas, black oil
Novokuibyshavskaya CHP-2	295	867	1954-1962	2002	n/a
Syzranskaya CHP	160	713	1947	1987	shale
Saratovskaya CHP-1	18	208	1930	1970	gas, black oil
Saratovskaya CPP	54	478	1930-1951	1991	gas, black oil
Saratovskaya CHP-2	226	818	1955	1995	gas, black oil
Engelskaya CHP-3	180	724	1958-1962	2002	gas, black oil
Balakovskaya CHP-4	365	1232	1962	2002	gas, black oil
Saratovskaya CHP-5	440	1240	1976	2016	gas, black oil
Саратовская ГРЭС	54	506	1926	1966	gas
Ulyanovskaya CHP-1	435	1424	1947	1987	n/a
Ulyanovskaya CHP-2	417	1216	1978	2018	n/a
Ulyanovskaya CHP-3	-	500	1989	2029	n/a
Kargalinskaya CHP	320	1030			n/a
Sakmarskaya CHP	460	1576	1969	2009	n/a
Orskaya CHP	245	1349	1938	1978	n/a
Mednogorskaya CHP	14	100	1955, 1956, 1992	2032	n/a
TGC-9 (TGC-9, 2012)					
Bereznikovskaya CHP-2	119.5	645.8	1947	1987	gas, black oil
Kizelovskaya CPP-3	26	231.9	1924, 2004	2044	gas
Bereznikovskaya CHP-4	29.2	276.4	1931	1971	gas, black oil
Bereznikovskaya CHP-10	33	405	1953	1993	gas, black oil
Zakamskaya CHP-5	81	640.5	1936	1976	gas
Permskaya CHP-6	56.7	1233.5	1942, 1944, 1948	1988	gas, black oil
Permskaya CHP-9	447	1633.8	1957	1997	gas, black oil, coal
Permskaya CHP-13	18	332.2	1959.2004	2044	gas, black oil
Permskaya CHP-14	330	941	1966-1967	2007	n/a
Chaikovskaya CHP-18	200	466	1978	2018	gas, black oil, coal
Nizhneturinskaya CPP	284	510	1950	1990	coal
Novo-Cserdlovskaya CPP	550	890	1982	2022	gas
Bogoslovskaya CHP	141	1045	1951	1991	n/a
Kachkanarskaya CHP	50	352	1963	2003	n/a
Krasnogorskaya CHP	121	1006	1935-1966	2006	gas, coal
Pervouralskaya CHP	36	967	1956	1996	gas

Name of the station	Installed electricity capacity, MW	Installed heat capacity Gcal/h	Year of construction, reconstruction	Expire date	Type of fuel by relevancy[62]
Sverdlovskaya CHP	36	1430	1935-1980	2020	gas
Borkutinskaya CHP-1	25	176	1942	1982	coal, black oil
Borkutinskaya CHP-2	270	429	1955	1995	coal
Intinskaya CHP	18	171	1949	1989	n/a
Sosnogorskaya CHP	377	313	1960-1963	2003	n/a
TGC-10 (Fortum) (TGC-10, 2012)					
Tyuminslaya CHP-1	662	1897	1960-2011	2051	gas, coal
Tyuminslaya CHP-2	755	1640	1986	2026	gas, coal
Tobolskaya CHP	665	2808	1980-2011	2051	gas, coal
Chelyabinskaya CHP-1	149	1560	1942	1982	gas, coal
Chelyabinskaya CHP-2	320	1112	1962	2002	gas, coal
Chelyabinskaya CHP-3	577	1326	1996-2011	2051	gas, coal
Chelyabinskaya CPP	82	947	1930-2007	2047	gas, coal
Argayashskaya CHP	194	670	1954	1994	gas, coal
Nyaganskaya CPP			under construction		
TGC-11 (TGC-11, 2012)					
Omskaya CHP-2	-	378	1941	1981	gas, coal, black oil
Omskaya CHP-3	350	1430	1954	1994	gas, black oil
Omskaya CHP-4	535	1500	1965	2005	coal, gas, black oil
Omskaya CHP-5	695	1735	1980	2020	coal, black oil
Tomskaya CPP-2	331	815	1945	1985	gas, coal, black oil
Tomskaya CHP-3	140	780	1988	2028	gas, black oil
TGC-12 (TGC-12, 2011)					
Tom-Usinskaya CPP	1272	278	1958	1998	coal, black oil
Belovskaya CPP	1200	272	1964	2004	coal, black oil
Kemerovskaya CPP	485	1540	1934	1974	coal, gas, black oil
Kemerovskaya CHP	85		1929	1969	n/a
Novo-Kemerovskaya CHP	565	1449	1955	1995	coal, gas, black oil
Kuznezkaya CHP	108	976	1944	1984	coal, gas, black oil
Barnaulskaya CHP-2	339	1371	1955	1995	n/a
Barnaulskaya CHP-3	430	1462	1976	2016	coal, gas, black oil
Barnaulskaya CHP-1	20.2	170	1936	1976	gas, coal
TGC-13 (TGC-13, 2012)					
Krasnoyarskaya CHP-1	481	1677	1943-1975	2015	coal, black oil
Krasnoyarskaya CHP-2	465	1405	1979	2019	coal, black oil
Krasnoyarskaya CHP-3	208	631.5	1992-2012	2052	coal, black oil
Nazarovskaya CPP	1210	870	1961	2001	coal, black oil
Minusinskaya CHP	85	330.4	1997-2011	2051	coal, black oil
Kanskaya CHP	24	325	1953	1993	coal, black oil
Krasnoyarskaya CHP-4	-	486	2009	2049	coal, black oil
Abakanskaya CHP	270	625	1982	2022	coal, black oil
TGC-14 (TGC-14, 2012)					
Chitinskaya CHP-1	471	987	1965	2005	coal, black oil
Chitinskaya CHP-2	12	229	1936	1976	coal, black oil

Name of the station	Installed electricity capacity, MW	Installed heat capacity Gcal/h	Year of construction, reconstruction	Expire date	Type of fuel by relevancy[62]
Sherlovogorskaya CHP	12	99	1956	1996	coal, black oil
Priargunskaya CHP	24	110	1961	2001	coal, black oil
Ulan-Udenskaya CHP-1	148.77	852	1936	1976	coal, black oil
Ulan-Udenskaya CHP-2	-	380	1991	2031	coal, black oil
Timlyuiskaya CHP	-	59.6	1953	1993	coal, black oil
Far East Generation Company (DVGK, 2012)					
Chabarovskaya CHP-1	435	1200.20	1954-1972	2012	coal, gas, black oil
Chabarovskaya CHP-3	720	1640	1985	2025	coal, gas
Komsomolskaya CHP-2	240.5	786	1934-1939	1979	coal, gas
Komsomolskaya CHP-3	360	1240	1988, 2012	2052	coal, gas
Amurskaya CHP-1	285	1169	1956	1996	coal, gas
Maiskaya CPP	80.85	15.40	1938	1978	coal, gas
Nikolaevskaya CHP	130.60	321.20	1973-1975	2015	coal, gas
Vladivostokskaya CHP-2	497	1051	1970	2010	coal, gas, black oil
Vladivostokskaya CHP-1	n/a	-	1911	1951	n/a
Birobidzhanskaya CHP	n/a	338	1958	1998	coal
Artemovskaya CHP	400	297	1936	1976	coal, gas
Partizanskaya CPP	203	160	1954	1994	coal, gas
Primorskaya CPP	1 467	237	1974	2014	coal, gas
Blagovesh'enskaya CPP	280	817	1976	2016	coal, gas
Raichihinskaya CPP	102	238.10	1953	1993	coal, gas
Neryungrinskaya CPP	570	1220	1983-1985	2025	coal, gas
Chul'manskaya CPP	48	165	1962	2002	coal, gas

Appendix 3: Calculations.

Table 21: Technical parameters of new conventional PPs construction in North-West IES for the period 2011-2017 (SO, 2011)

North-West IES								
	2011	**2012**	**2013**	**2014**	**2015**	**2016**	**2017**	**Total**
Nuclear PP, MW				**1176.0**		**2326.0**		**3502.0**
Heat PP, MW:	**1350.0**	**630.0**	**440.0**	**657.8**	**70.0**	**199.0**	**244.0**	**3590.8**
CHP	650.0	630.0	440.0	657.8	70.0	199.0	244.0	2890.8
CPP	700.0							700.0
Investment costs Nuclear, bn euro	0.0	0.0	0.0	2.4	0.0	4.7	0.0	7.0
Investment costs Heat PP, bn euro	1.6	0.8	0.5	0.8	0.1	0.2	0.3	4.4
Total investment costs, bn euro	**1.6**	**0.8**	**0.5**	**3.2**	**0.1**	**4.9**	**0.3**	**11.4**
Full load hour Nuclear PP	6964.0	7488.0	7349.0	6738.0	7285.0	5747.0	6060.0	
Full load hours Heat PP	3882.0	3201.0	3299.0	3143.0	3168.0	3136.0	3200.0	
Nuclear PP, TWh	0.0	0.0	0.0	7.9	0.0	13.4	0.0	21.3
Heat PP, TWh:	5.2	2.0	1.5	2.1	0.2	0.6	0.8	12.4
CHP	2.5	2.0	1.5	2.1	0.2	0.6	0.8	9.7
CPP	2.7	0.0	0.0	0.0	0.0	0.0	0.0	2.7

Table 22: Replacement of new conventional PPs by RE technologies in North-West IES, MW (Author)

Type of RE technology	Type of PP to exchange		
	NuclearPP	CHP	CPP
Biomass			
cost CHP		16.1	
cost basic	2326.0		
Biogas, Agric			
cost CHP			
cost basic			
Biogas, Landfill			
cost CHP		783.7	
cost basic	1176.0		
Biogas, Sewage			
cost CHP		489.5	
cost basic			
Hydro, 20 MW			
Hydro, 9.5 MW		1601.5	700.0
Wind			
Solar			
Total, MW	**3502.0**	**2890.8**	**700.0**
Total North-West IES, MW			**7092.8**

Table 23: Investment cost for RE technologies construction in North-West IES, bn euro (Author)

Type of RE technology	Type of PP to exchange		
	Nuclear PP	CHP	CPP
Biomass			
cost CHP		0.1	
cost basic	6.1		
Biogas: Agriculture			
cost CHP			
cost basic			
Biogas: Landfill			
cost CHP		1.4	
cost basic	1.9		
Biogas: Sewage			
cost CHP		1.5	
cost basic			
Hydro, 20 MW,			
Hydro, 9.5 MW		2.1	0.9
Wind			
Solar			
Total	**8.0**	**5.0**	**0.9**
Total North-West IES, bn euro			**13.9**

Table 24: Technical parameters of new conventional PPs construction in Central IES for the period 2011-2017 (SO, 2011)

Central IES								
	2011	**2012**	**2013**	**2014**	**2015**	**2016**	**2017**	**Total**
Nuclear PP, MW	**1000.0**		**1180.0**		**1180.0**			**3360.0**
Heat PP, MW:	**1149.0**	**737.6**	**313.8**	**2183.8**	**836.0**	**60.0**		**5280.2**
CHP	1059.0	412.6	50.0	1970.0	416.0	60.0		3967.6
CPP	90.0	325.0	263.8	213.8	420.0			1312.6
Investment costs Nuclear, bn euro	2.0	0.0	2.4	0.0	2.4	0.0	0.0	6.8
Investment costs Heat PP, bn euro	1.4	0.9	0.4	2.6	1.0	0.1	0.0	6.4
Total investment costs, bn euro	**3.4**	**0.9**	**2.8**	**2.6**	**3.4**	**0.1**	**0.0**	**13.2**
Full load hour Nuclear PP	6428.0	7036.0	6865.0	7500.0	7286.0	7500.0	7500.0	
Full load hours Heat PP	4075.0	4105.0	4124.0	3700.0	3622.0	3789.0	3908.0	
Nuclear PP, TWh	6.4	0.0	8.1	0.0	8.6	0.0	0.0	23.1
Heat PP, TWh:	4.7	3.0	1.3	8.1	3.0	0.2	0.0	20.3
CHP	4.3	1.7	0.2	7.3	1.5	0.2	0.0	15.2
CPP	0.4	1.3	1.1	0.8	1.5	0.0	0.0	5.1

Table 25: Replacement of new conventional PPs by RE technologies in Central IES, MW (Author)

Type of RE technology	Type of PP to exchange		
	NuclearPP	**CHP**	**CPP**
Biomass			
cost CHP		3029	
cost basic			213.8
Biogas, Agriculture			
cost CHP			
cost basic			
Biogas, Landfill			
cost CHP			
cost basic	3360		1098.8
Biogas, Sewage			
cost CHP		522.6	
cost basic			
Hydro, 20 MW			
Hydro, 9.5 MW		416	
Wind			
Solar			
Total, MW	**3360**	**3967.6**	**1312.6**
Total Central IES, MW			**8640.2**

Table 26: Investment cost for RE technologies construction in Central IES, bn euro (Author)

Type of RE technology		Type of PP to exchange		
		NuclearPP	CHP	CPP
Biomass				
	cost CHP		10.56	
	cost basic			0.56
Biogas, Agriculture				
	cost CHP			
	cost basic			
Biogas, Landfill				
	cost CHP			
	cost basic	5.54		1.81
Biogas, Sewage				
	cost CHP		1.55	
	cost basic			
Hydro, 20 MW				
Hydro, 9.5 MW			0.54	
Wind				
Solar				
Total		**5.54**	**12.65**	**2.37**
Total Central IES, bn euro				**20.57**

Table 27: Technical parameters of new conventional PPs construction in VolgaIES for the period 2011-2017 (SO, 2011)

Volga IES								
	2011	**2012**	**2013**	**2014**	**2015**	**2016**	**2017**	**Total**
Nuclear PP, MW								**0.0**
Heat PP, MW:	**261.0**	**121.0**	**690.0**	**845.0**	**458.0**	**951.0**	**240.0**	**3566.0**
CHP	261.0	121.0	690.0	690.0	128.0	951.0	240.0	3081.0
CPP				155.0	330.0			485.0
Investment costs Nuclear, bn euro	0.0	0.0	0.0	0.0	0.0	0.0	0.0	0.0
Investment costs Heat PP, bn euro	0.3	0.1	0.8	1.0	0.6	1.2	0.3	4.3
Total investment costs, bn euro	**0.3**	**0.1**	**0.8**	**1.0**	**0.6**	**1.2**	**0.3**	**4.3**
Full load hour Nuclear PP	7466.0	7392.0	7539.0	7417.0	7392.0	7490.0	7490.0	
Full load hours Heat PP	3776.0	3932.0	3936.0	3895.0	3895.0	3923.0	4033.0	
Nuclear PP, TWh	0.0	0.0	0.0	0.0	0.0	0.0	0.0	0.0
Heat PP, TWh:	1.0	0.5	2.7	3.3	1.8	3.7	1.0	14.0
CHP	1.0	0.5	2.7	2.7	0.5	3.7	1.0	12.1
CPP	0.0	0.0	0.0	0.6	1.3	0.0	0.0	1.9

Table 28: Replacement of new conventional PPs by RE technologies in Volga IES, MW (Author)

Type of RE technology	Type of PP to exchange		
	NuclearPP	CHP	CPP
Biomass			
cost CHP			
cost basic			
Biogas, Agriculture			
cost CHP			
cost basic			
Biogas, Landfill			
cost CHP		1641.0	
cost basic		27.0	
Biogas, Sewage			
cost CHP			
cost basic			
Hydro, 20 MW			
Hydro, 9.5 MW		1595.0	
Wind			303.0
Solar			
Total, MW	**0.0**	**3263.0**	**303.0**
Total Volga IES, MW			**3565.9**

Table 29: Investment cost for RE technologies construction in Volga IES, bn euro (Author)

Type of RE technology	Type of PP to exchange		
	NuclearPP	CHP	CPP
Biomass			
cost CHP			
cost basic			
Biogas, Agriculture			
cost CHP			
cost basic			
Biogas, Landfill			
cost CHP		3.0	
cost basic		0.0	
Biogas, Sewage			
cost CHP			
cost basic			
Hydro, 20 MW			
Hydro, 9.5 MW		2.1	
Wind			0.4
Solar			
Total	**0.0**	**5.1**	**0.4**
Total Volga IES, bn euro			**5.5**

Table 30: Technical parameters of new conventional PPs construction in South IES for the period 2011-2017 (SO, 2011)

South IES								
	2011	2012	2013	2014	2015	2016	2017	Total
Nuclear PP, MW				**1070.0**			**1070.0**	**2140.0**
Heat PP, MW:	**1110.0**	**385.0**	**690.0**	**840.0**	**700.0**	**420.0**	**104.0**	**4249.0**
CHP	710.0	385.0	370.0				104.0	1569.0
CPP	400.0		320.0	840.0	700.0	420.0		2680.0
Investment costs Nuclear, bn euro	0.0	0.0	0.0	2.2	0.0	0.0	2.2	4.3
Investment costs Heat PP, bn euro	1.3	0.5	0.8	1.0	0.8	0.5	0.1	5.2
Total investment costs, bn euro	**1.3**	**0.5**	**0.8**	**3.2**	**0.8**	**0.5**	**2.3**	**9.5**
Full load hour Nuclear PP	7600.0	7350.0	7600.0	7296.0	7590.0	7590.0	7464.0	
Full load hours Heat PP	4119.0	4155.0	4194.0	3917.0	3893.0	3893.0	3740.0	
Nuclear PP, TWh	0.0	0.0	0.0	7.8	0.0	0.0	8.0	15.8
Heat PP, TWh:	4.6	1.6	2.9	3.3	2.7	1.6	0.4	17.1
CHP	2.9	1.6	1.6	0.0	0.0	0.0	0.4	6.5
CPP	1.6	0.0	1.3	3.3	2.7	1.6	0.0	10.6

Table 31: Replacement of new conventional PPs by RE technologies in South IES, MW (Author)

Type of RE technology	Type of PP to exchange		
	NuclearPP	CHP	CPP
Biomass			
cost CHP	710.0		
cost basic	1070.0		
Biogas, Agriculture			
cost CHP			
cost basic			
Biogas, Landfill			
cost CHP		646.4	
cost basic	1070.0		320.0
Biogas, Sewage			
cost CHP			212.2
cost basic			
Hydro, 20 MW			
Hydro, 9.5 MW			2360.0
Wind			
Solar			
Total, MW	**2850.0**	**646.4**	**2892.2**
Total South IES, MW			**6388.6**

Table 32: Investment cost for RE technologies construction in South IES, bn euro (Author)

Type of RE technology	Type of PP to exchange		
	NuclearPP	CHP	CPP
Biomass			
cost CHP	1.9		
cost basic	3.7		
Biogas, Agriculture			
cost CHP			
cost basic			
Biogas, Landfill			
cost CHP		1.2	
cost basic	1.8		0.5
Biogas, Sewage			
cost CHP			0.6
cost basic			
Hydro, 20 MW			
Hydro, 9.5 MW			3.0
Wind			
Solar			
Total	**7.4**	**1.2**	**4.2**
Total South IES, bn euro			**12.7**

Table 33: Technical parameters of new conventional PPs construction in Ural IES for the period 2011-2017 (SO, 2011)

Ural IES								
	2011	2012	2013	2014	2015	2016	2017	Total
Nuclear PP, MW				**880.0**				**880.0**
Heat PP, MW:	**2858.1**	**1088.0**	**2143.0**	**1565.0**	**3853.0**	**533.0**	**594.0**	**12634.1**
CHP	1397.0	220.0	915.0	295.0	955.0	73.0	594.0	4449.0
CPP	1461.1	868.0	1228.0	1270.0	2898.0	460.0		8185.1
Investment costs Nuclear, bn euro	0.0	0.0	0.0	1.8	0.0	0.0	0.0	1.8
Investment costs Heat PP, bn euro	3.5	1.3	2.6	1.9	4.7	0.6	0.7	15.3
Total investment costs, bn euro	**3.5**	**1.3**	**2.6**	**3.7**	**4.7**	**0.6**	**0.7**	**17.1**
Full load hour Nuclear PP	6500.0	7167.0	7167.0	6757.0	7500.0	7500.0	7500.0	
Full load hours Heat PP	5577.0	5524.0	5467.0	5296.0	5050.0	5110.0	5169.0	
Nuclear PP, TWh	0.0	0.0	0.0	5.9	0.0	0.0	0.0	5.9
Heat PP, TWh:	15.9	6.0	11.7	8.3	19.5	2.7	3.1	67.2
CHP	7.8	1.2	5.0	1.6	4.8	0.4	3.1	23.8
CPP	8.1	4.8	6.7	6.7	14.6	2.4	0.0	43.4

Table 34: Replacement of new conventional PPs by RE technologies in Ural IES, MW (Author)

Type of RE technology	Type of PP to exchange		
	NuclearPP	CHP	CPP
Biomass			
cost CHP		2579.0	259.6
cost basic			
Biogas, Agriculture			
cost CHP			
cost basic			3106.2
Biogas, Landfill			
cost CHP		1870.0	
cost basic			
Biogas, Sewage			
cost CHP			
cost basic			
Hydro, 20 MW			
Hydro, 9.5 MW	880.0		3370.6
Wind			1448.8
Solar			
Total, MW	**880.0**	**4449.0**	**8185.2**
Total Ural IES, MW			**13514.2**

Table 35: Investment cost for RE technologies construction in Ural IES, bn euro (Author)

Type of RE technology	Type of PP to exchange		
	NuclearPP	CHP	CPP
Biomass			
cost CHP		9.0	0.7
cost basic			
Biogas, Agriculture			
cost CHP			
cost basic			10.6
Biogas, Landfill			
cost CHP		3.4	
cost basic			
Biogas, Sewage			
cost CHP			
cost basic			
Hydro, 20 MW			
Hydro, 9.5 MW	1.1		4.3
Wind			1.9
Solar			
Total	**1.1**	**12.4**	**17.6**
Total Ural IES, bn euro			**31.1**

Table 36: Technical parameters of new conventional PPs construction in Siberia IES for the period 2011-2017 (SO, 2011)

Siberia IES								
	2011	**2012**	**2013**	**2014**	**2015**	**2016**	**2017**	**Total**
Nuclear PP, MW								**0.0**
Heat PP, MW:	**12.0**	**572.8**	**112.0**	**1315.0**	**1397.0**	**730.0**	**260.0**	**4398.8**
CHP		359.0	112.0	455.0	287.0		100.0	1313.0
CPP	12.0	213.8		860.0	1110.0	730.0	160.0	3085.8
Investment costs Nuclear, bn euro	0.0	0.0	0.0	0.0	0.0	0.0	0.0	0.0
Investment costs Heat PP, bn euro	0.01	0.7	0.1	1.6	1.7	0.9	0.3	5.3
Total investment costs, bn euro	**0.01**	**0.7**	**0.1**	**1.6**	**1.7**	**0.9**	**0.3**	**5.3**
Full load hour Nuclear PP								
Full load hours Heat PP	4915.0	4852.0	4729.0	4644.0	4768.0	4876.0	5040.0	
Nuclear PP, TWh	0.0	0.0	0.0	0.0	0.0	0.0	0.0	0.0
Heat PP, TWh:	0.1	2.8	0.5	6.1	6.7	3.6	1.3	21.0
CHP	0.0	1.7	0.5	2.1	1.4	0.0	0.5	6.3
CPP	0.1	1.0	0.0	4.0	5.3	3.6	0.8	14.7

Table 37: Replacement of new conventional PPs by RE technologies in Siberia IES, MW (Author)

Type of RE technology	Type of PP to exchange		
	Nuclear PP	CHP	CPP
Biomass			
cost CHP			
cost basic			
Biogas, Agriculture			
cost CHP			
cost basic			
Biogas, Landfill			
cost CHP		1313	
cost basic			
Biogas, Sewage			
cost CHP			
cost basic			
Hydro, 20 MW			
Hydro, 9.5 MW			3085.8
Wind			
Solar			
Total, MW	**0**	**1313**	**3085.8**
Total Siberia IES, MW			**4398.8**

Table 38: Investment cost for RE technologies construction in Siberia IES, bn euro (Author)

Type of RE technology	Type of PP to exchange		
	Nuclear PP	CHP	CPP
Biomass			
cost CHP			
cost basic			
Biogas, Agriculture			
cost CHP			
cost basic			
Biogas, Landfill			
cost CHP		2.4	
cost basic			
Biogas, Sewage			
cost CHP			
cost basic			
Hydro, 20 MW			
Hydro, 9.5 MW			4.0
Wind			
Solar			
Total	**0.0**	**2.4**	**4.0**
Total Siberia IES, bn euro			**6.3**

Table 39: Technical parameters of new conventional PPs construction in Far East IES for the period 2011-2017 (SO, 2011)

East IES								
	2011	**2012**	**2013**	**2014**	**2015**	**2016**	**2017**	**Total**
Nuclear PP, MW								**0.0**
Heat PP, MW:			**352.5**	**110.0**	**185.0**		**70.0**	**717.5**
CHP			352.5	110.0	185.0		70.0	717.5
CPP								0.0
Investment costs Nuclear, bn euro	0.0	0.0	0.0	0.0	0.0	0.0	0.0	0.0
Investment costs Heat PP, bn euro	0.00	0.0	0.4	0.1	0.2	0.0	0.1	0.9
Total investment costs, bn euro	**0.00**	**0.00**	**0.43**	**0.13**	**0.22**	**0.00**	**0.08**	**0.9**
Full load hour Nuclear PP								
Full load hours Heat PP	3871.0	3492.0	3678.0	3811.0	3739.0	3898.0	4172.0	
Nuclear PP, TWh	0.0	0.0	0.0	0.0	0.0	0.0	0.0	0.0
Heat PP, TWh:	0.0	0.0	1.3	0.4	0.7	0.0	0.3	2.7
CHP	0.0	0.0	1.3	0.4	0.7	0.0	0.3	2.7
CPP	0.0	0.0	0.0	0.0	0.0	0.0	0.0	0.0

Table 40: Replacement of new conventional PPs by RE technologies in Far East IES, MW (Author)

Type of RE technology	Type of PP to exchange		
	NuclearPP	CHP	CPP
Biomass			
cost CHP			
cost basic			
Biogas, Agriculture			
cost CHP			
cost basic			
Biogas, Landfill			
cost CHP		717.5	
cost basic			
Biogas, Sewage			
cost CHP			
cost basic			
Hydro, 20 MW			
Hydro, 9.5 MW			
Wind			
Solar			
Total, MW	**0**	**717.5**	**0**
Total Far East IES, MW			**717.5**

Table 41: Investment cost for RE technologies construction in Far East IES, bn euro (Author)

Type of RE technology	Type of PP to exchange		
	NuclearPP	CHP	CPP
Biomass			
cost CHP			
cost basic			
Biogas, Agriculture			
cost CHP			
cost basic			
Biogas, Landfill			
cost CHP		1.3	
cost basic			
Biogas, Sewage			
cost CHP			
cost basic			
Hydro, 20 MW			
Hydro, 9.5 MW			
Wind			
Solar			
Total	**0.0**	**1.3**	**0.0**
Total Far East IES, bn euro			**1.3**